FEARFUL SYMMETRY

The Search for Beauty
in Modern Physics

With a new foreword by Roger Penrose

A. Zee

PRINCETON UNIVERSITY PRESS
Princeton and Oxford

Published by Princeton University Press, 41 William Street,
Princeton, New Jersey 08540

First published by Macmillan Publishing Company, 1986
First Princeton edition, for the Princeton Science Library, 1999
Reprinted, with a new foreword, 2007
ISBN-13: 978-0-691-13482-6

The Library of Congress has cataloged the first Princeton edition as follows

Zee, A.
    Fearful symmetry : the search for beauty in modern physics / A. Zee.
    p.   cm.
    With a new preface and afterword.
    Originally published: New York : Macmillan Pub. Co., 1986
    Includes bibliographical references and index.
    ISBN 0-691-00946-5 (pb. : alk. paper)
    1. Symmetry (Physics) 2. Physics—Philosophy. 3. Renormalization
    (Physics) I. Title.
    QC174.17.S9Z4   1999
    530.1—dc21   99–44882

Printed on acid-free paper. ∞

press.princeton.edu

Printed in the United States of America

10   9   8   7   6   5   4   3

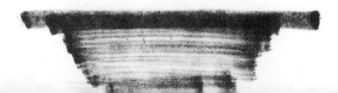

To Gretchen

Tyger! Tyger! burning bright
In the forests of the night,
What immortal hand or eye
Could frame thy fearful symmetry?

—William Blake
(From *Poems of William Blake*,
edited by W. B. Yeats. London:
G. Routledge & Sons, 1905.)

Contents

Foreword

The notion of *symmetry* has manifestations that are wide-spread in human culture. Symmetry also exhibits itself in a great many features of the operation of the Natural world. It is a notion that is simple enough to be understood and made use of by a young child; yet it is subtle enough to be central to our deepest and most successful physical theories describing the inner workings of Nature. Symmetry is thus a concept that is simultaneously obvious and profound.

Symmetry has many areas of application. Some are basically practical, such as those made use of in engineering—for example, with the bilateral symmetry of most aeroplane designs or in the symmetry in bridge construction—or, at a more mundane level, in the satisfactory creation of furniture. But other uses of symmetry are more evidently purely aesthetic, and can be central to many artistic creations, where it may provide a key ingredient underlying the sublime beauty of various great works of art. Moreover, this is true over a considerable range of disciplines, such as in painting or sculpture, or in music or literature. In mathematics, the notion of symmetry is the starting point of vast areas of deep theory, providing much penetrating insight of enormous scope. In the science of crystallography it is crucial, as it is also in many aspects of chemistry. It is evident to the eye that symmetry is also important to biological function. There is great symmetry to be found throughout the plant and animal kingdoms, and this symmetry can contribute in many different ways to the efficient functioning of an organism. Bilateral symmetry, for example, is almost universal among animals. Yet it is occasionally grossly violated, such as in the twisting spiral shape of a snail's shell. But with such a shell, we find that there emerges a deeper type of symmetry, not evident at first: a symmetry under rotational motion, provided that this is accompanied by uniform expansion or contraction.

It is in modern fundamental physics, though, that we find the most subtle interplay between symmetry and asymmetry, for, as twentieth-century physics has revealed, there is a special role for symmetry in Nature's basic forces that is both central and sometimes enigmatically violated. Some of the underlying ideas are quite simple to grasp, but others, distinctly sophisticated. However, I believe that this undoubted sophistication should not be held as a reason for denying the lay public an opportunity to access some of these ideas, which often exhibit a kind of sublime beauty. Yet, it is not so easy to present such ideas in a way accessible to a lay reader, while at the same time holding the reader's interest and conveying some nontrivial understanding. Tony Zee, however, achieves this in a masterful way in his classic, here reprinted, *Fearful Symmetry: The Search for Beauty in Modern Physics*.

Fundamental physics is genuinely difficult, and the crucial role of symmetry is often a subtle one. For this good reason it is not often satisfactorily explained to a lay audience at length or in significant depth. In the arts, on the other hand, it is usually not hard to produce accessible works fully extolling the use of various qualities of symmetry, often merely by use of a good picture, as is the case also in architecture and engineering. This seems to be true as well in biological descriptions and in books depicting the natural beauty of crystalline substances. There are even many popular works in which the symmetrical forms of pure geometry, or of other areas of pure mathematics, can be made very accessible even to the mathematically unsophisticated reader. But there is something especially hard to communicate about the roles of symmetry in basic physics, particularly because these roles are often very abstract and they frequently depend upon the highly nonintuitive and confusing fundamental principles of *quantum mechanics*. The demands of quantum physics are themselves not easy to comprehend. They are often subtle, and have a special importance to the workings of Nature at the deepest levels that human understanding has yet been able to penetrate. It is when the ideas of symmetry and quantum mechanics come together that we find the special subtleties that are crucial to our modern understanding of the basic forces of Nature.

All this serves to emphasize the difficulties that confronted Zee's task, and to enhance our admiration for his superb achievement. But there is one thing to his advantage that he was able to exploit: the fact that symmetry and *beauty* are closely linked. By emphasizing the beauty in the use of symmetry that lies behind our modern theories of the forces of Nature, rather than dwelling on the mathematical details, he is able to side-step most of the technicality and many of the deep remaining difficulties that presently confound further progress. This is a story that, indeed, has not reached its end, as it is still beset with enigmas and apparent inconsistencies. Of course, this makes it all the more fascinating to study the progress that has been made so far, and perhaps to entice more readers to enter, in a serious way, this wonderful world—a world that that they may not have considered before.

And for this, Tony Zee here serves as a superb guide!

Roger Penrose
March 2007

Preface 1999

I am happy that I decided to write *Fearful Symmetry*. During a visit to the University of Texas in 1984, I was chatting with the eminent physicist Steve Weinberg when his secretary brought him his mail, which happened to contain a review of his second popular physics book. Our conversation naturally turned to writing popular physics books. Various physicists had encouraged me to write a textbook, on quantum field theory in particular. Not only did Weinberg encourage me to write a popular physics book instead, he also introduced me to his publisher and offered valuable advice. Thus, a few months later I found myself having lunch in New York with Weinberg's publisher. I was told to bring a sample chapter, which I chose to devote to conservation laws; it started with the sentence "There is no free lunch." The publisher laughed and said, "There is." Since then I have been taken to lunch and dinner by various publishers, editors, and agents, and I have also re-learned that in some sense there is indeed no free lunch. The curious reader will find the sentence about free lunch in Chapter 8, although it no longer starts the chapter.

I am happy that I wrote *Fearful Symmetry* because of all the wonderful letters (and even presents) I have received from appreciative readers, because of all the ego-warming reviews, because of the pleasure of hearing my words read by a professional actor in a Library of Congress audio tape for the blind, and because of seeing my immortal prose translated into several foreign languages but, most of all, I am happy because the book allowed me to get out of the physics community once in a while. I was invited to lecture about symmetry at all sorts of interesting places, for example at the National Center for the Performing Arts in Bombay, where I learned about symmetry in classical Indian dances, and at the Berlin Academy of Arts, where I was invited to participate in an international symposium on racism. (What racism has to do with symmetry is not clear to me.) *Fearful Symmetry* launched my career as a writer. The pages I cut out of my manuscript became the core of my second popular physics book *An Old Man's Toy*. Before long, I found parts of *Fearful Symmetry* excerpted in a college textbook on writing and received invitations to speak to classes on creative writing.

I am not happy that Macmillan Publishing Company allowed my book to go out of print. I have subsequently been acquainted with some horrifying statistics about the commercial publishing industry. So I am truly happy, particularly as an alumnus of Princeton University, that Princeton University Press is publishing a new edition of *Fearful Symmetry*. I thank the physicists Murph Goldberger, Dave Spergel, and Sam Treiman for encouraging this project, and my editors Trevor Lipscombe and Donna Kronemeyer for their good work.

I would like to thank Joe Polchinski and Roger Shepard for reading the Afterword added for this Princeton University edition. I think it appropriate to thank once again the physicists who read parts or all of the original manuscript: Bill Bialek, Sidney Coleman, Murray Gell-Mann, Tsung-dao Lee, Heinz Pagels, Steve Weinberg, and Frank Wilczek. Finally, I must thank Gretchen Zee for her years of support and love.

Santa Barbara
March 1999

Preface

In *Fearful Symmetry*, I wish to discuss the aesthetic motivations that animate twentieth-century physics. I am interested more in conveying to the reader a sense of the intellectual framework within which fundamental physics operates and less in explaining the factual content of modern physics.

Albert Einstein once said, "I want to know how God created this world. I am not interested in this or that phenomenon, in the spectrum of this and that element. I want to know His thoughts, the rest are details."

As a physicist, I am much enamored of the sentiment expressed by Einstein. While the vast majority of contemporary physicists are engaged in explaining specific phenomena, and rightly so, a small group, the intellectual descendants of Einstein, have become more ambitious. They have entered the forest of the night in search of the fundamental design of Nature and, in their limitless hubris, have claimed to have glimpsed it.

Two great principles guide this search: symmetry and renormalizability. Renormalizability refers to how physical processes with different characteristic lengths are related to each other. While I will touch on renormalizability, my focus will be on symmetry as the unifying aesthetic viewpoint through which fundamental physicists look at Nature.

There has been a growing interest in modern physics over the last few years. Expositions of the "new" physics abound. By now many of us have learned that there are billions and billions of galaxies, each containing billions and billions of stars. We have been told that the world may be understood in terms of subnuclear particles, most of which live for a billionth of a billionth of a second. The informed reader has been astounded and dazzled. Yes, indeed, the world of modern physics is wonderfully bizarre. Particles carrying Greek names jitterbug to the music of the quantum in defiance of classical determinism. But ultimately, the reader may come away with a sense of being fed simply the facts which, while truly amazing, become tiresome of themselves.

This book is addressed to the intellectually curious reader who wants to go beyond the facts. I have a mental image of that reader: someone I once knew in my youth; someone who may have since become an architect, an artist, a dancer, a stockbroker, a biologist, or a lawyer; someone who is interested in the intellectual and aesthetic framework within which fundamental physicists operate.

This does not mean that the astounding discoveries of modern physics will not be explained in this book. They will have to be explained before I can meaningfully discuss the intellectual framework of modern physics. I hope, however, that the reader will come away with not only a nodding acquaintance with certain astounding facts, but also with a sense of the framework without which they would remain, simply, facts.

I have not attempted to give a detailed and balanced history of symmetry in physics. Any account in which major developments are attributed to a handful of individuals cannot claim to be history, and any assertion to that effect must be rejected categorically. In speaking of certain developments in modern particle physics, the eminent physicist Shelley Glashow once remarked: "Tapestries are made by many artisans working together. The contributions of separate workers cannot be discerned in the completed work, and the loose and false threads have been covered over. So it is in our picture of particle physics. . . . [The standard theory] did not arise full blown in the mind of one physicist, nor even of three. It, too, is the result of the collective endeavor of many scientists, both experimenters and theorists." And yet, in a popular account such as this I am forced, inevitably, to simplify history. I trust that the reader understands.

—Santa Barbara, April 1986

Acknowledgments

First and foremost, I would like to thank my wife, Gretchen. Her incisive and critical comments, as well as her loving support, were essential. She would read each chapter as I went along, brutally slashing through the typescript. "I can't understand this!" she would scrawl across the page. And back to the writing-board I would go.

Our friends Kim Beeler, Chris Groesbeck, Martha and Frank Retman, and Diane Shuford—a psychologist, a student of art history, a lawyer, and two architects between them—read various portions of the manuscript to ensure that the text would be understandable to the lay reader.

Heinz Pagels and Steve Weinberg, two distinguished colleagues who have published popular books on physics, both encouraged me to pursue my idea of writing one about symmetry. They generously advised me on various aspects of writing and publishing, and introduced me to their friends in the publishing world.

I am grateful to Tsung-dao Lee, Heinz Pagels, and Steve Weinberg for reading the manuscript, and for their helpful and encouraging comments. I would also like to thank Sidney Coleman and Frank Wilczek for reading Chapter 12, Murray Gell-Mann for reading Chapter 11, and Bill Bialek for reading the galley proofs.

I am fortunate to have Charles Levine as my editor. His advice and support were indispensable. He was reassuring when I needed reassurance and critical when I needed criticism. I have come to value him as a friend.

My line editor, Catherine Shaw, clearly did a good job since I had to spend nearly two months rewriting the manuscript to answer all her comments. "I can't understand this!" she would exclaim in her turn. The book became clearer as a result. The manuscript was further smoothed by my copy editor, Roberta Frost.

Martin Kessler offered helpful advice during the early stages of this project.

I have also benefited from the advice of my agents, John Brockman and Katinka Matson.

The artists listed below helped to make the book more visually appealing and clearer.

I am pleased that the design director for the book is Helen Mills, whose brother Robert we will meet in Chapter 12. An appreciation for symmetry and balance appears to run in the family.

Finally, I would like to thank Debra Witmoyer, Lisa Lopez, Gwen Cattron, Katie Doremus, Karen Murphy, and Kresha Warnock for typing various portions of the manuscript.

FIGURE CREDITS

Bonnie Bright, figs. 3.4, 5.2, 6.3, 7.2, 7.3, 7.4, 10.2, 10.3, 11.1, 11.3, 12.1, 12.2, 12.3, 14.2, 15.2
Michael Cullen, figs. 3.5, 3.9, 9.1, 11.7, 13.2, 14.1, 14.4
Ji-jun Huang, fig. 15.1
Eric Junker, figs. 5.1, 5.3, 5.4
Joe Karl, figs. 2.1, 2.3, 4.2
Peggy Royster, figs. 4.3, 13.1
Clara Weis, fig. 4.1
Gretchen Zee, figs. 2.2, 7.1, 9.2, 10.1

I

SYMMETRY
AND DESIGN

1

In Search of Beauty

What I remember most clearly was that when I put down a suggestion that seemed to me cogent and reasonable, Einstein did not in the least contest this, but he only said, "Oh, how ugly." As soon as an equation seemed to him to be ugly, he really rather lost interest in it and could not understand why somebody else was willing to spend much time on it. He was quite convinced that beauty was a guiding principle in the search for important results in theoretical physics.
—H. Bondi

BEAUTY BEFORE TRUTH

My colleagues and I in fundamental physics are the intellectual descendants of Albert Einstein; we like to think that we too search for beauty. Some physics equations are so ugly that we cannot bear to look at them, let alone write them down. Certainly, the Ultimate Designer would use only beautiful equations in designing the universe! we proclaim. When presented with two alternative equations purporting to describe Nature, we always choose the one that appeals to our aesthetic sense. "Let us worry about beauty first, and truth will take care of itself!" Such is the rallying cry of fundamental physicists.

The reader may perhaps think of physics as a precise and predictive science and not as a subject fit for aesthetic contemplation. But, in fact, aesthetics has become a driving force in contemporary physics. Physicists have discovered something of wonder: Nature, at the fundamental level, is beautifully designed. It is this sense of wonder that I wish to share with you.

TRAINING OUR EYES

What is beauty? Philosophers pondering the meaning of aesthetics have produced weighty tomes, but an absolute definition of aesthetic values remains elusive. For one thing, fashion changes. The well-endowed ladies of Rubens no longer grace magazine covers. Aesthetic perceptions differ from culture to culture. Different conventions govern landscape painting in the East and West. The architectural designs of Bramante and I. M. Pei are beautiful in different ways. If there is no objective standard of beauty in the world of human creations, what system of aesthetics are we to use in speaking of the beauty of Nature? How are we to judge Nature's design?

In this book, I wish to explain how the aesthetic imperatives of contemporary physics make up a system of aesthetics that can be rigorously formulated. As my art history professors used to say, one has "to train one's eyes." To the architectural cognoscenti, the same principles that guide the Renaissance architect guide the postmodern. Likewise, physicists have to train their inner eye to see the universal principles guiding Nature's design.

INTRINSIC VERSUS EXTRINSIC BEAUTY

When I find a chambered nautilus at the seashore (or more likely in a shellshop), its beauty captivates me. But a developmental biologist would tell me that the perfect spiral is merely a consequence of unequal rate of shell growth. As a human being, I am no less enthralled by the beautiful nautilus knowing this fact, but as a physicist, I am driven to go beyond the extrinsic beauty that we can see. I want to discuss the beauty of neither the crashing wave nor the rainbow arcing across the sky, but the more profound beauty embodied in the physical laws that ultimately govern the behavior of water in its various forms.

LIVING IN A DESIGNER UNIVERSE

Physicists from Einstein on have been awed by the profound fact that, as we examine Nature on deeper and deeper levels, She appears ever more beautiful. Why should that be? We could have found ourselves living in an intrinsically ugly universe, a "chaotic world," as Einstein put it, "in no way graspable through thinking."

Musing along these lines often awakens feelings in physi-

Figure 1.1. *(Top)* Hokusai (1760–1849) "Mount Fuji Seen from Kanagawa."
(Courtesy Minneapolis Institute of Art)
(Bottom) Microphotograph of a snowflake *(R. B. Hoit, courtesy
Photo Researchers, Inc.)*
The beauty of water on two different levels.

cists best described as religious. In judging a physical theory purporting to describe the universe, Einstein would ask himself if he would have made the universe in that particular way, were he God. This faith in an underlying design has sustained fundamental physicists.

THE MUSIC VERSUS THE LIBRETTO

Popularizers of physics often regale us with descriptions of specific physical phenomena, astounding their readers with the fantastic discoveries of modern physics. I am more interested in conveying a sense of the intellectual and aesthetic framework of contemporary fundamental physics. Consider opera. The aficionado likes *Turandot*, but not primarily because of its libretto. The absurd story takes flight because of Puccini's music. On the other hand, it would be difficult to sit through an opera without knowing the story or worse yet, to listen only to the orchestral part. The music and libretto inform each other.

Similarly, to speak of the multitude of specific physical phenomena (the libretto) without placing them in the aesthetic framework of contemporary physics (the music) is boring and not particularly enlightening. I intend to give the reader the music of modern physics—the aesthetic imperatives that guide physicists. But just as an opera with the vocal part taken out would be senseless, a discussion of aesthetics without reference to actual physical phenomena is sterile. I will have to go through the libretto of physics. Ultimately, however, both as a fundamental physicist and as an opera lover, I must confess that my heart lies more with the music, and not the libretto.

LOCAL ORDINANCES VERSUS CONSTITUTIONAL PRINCIPLES

In a book about physics, the much-abused phrase "physical law" is certain to be bandied about. In civil law, one distinguishes between local ordinances and constitutional principles. So too in physics, there are laws and there are laws. Consider Hooke's law, stating that the force required to stretch a metal spring is proportional to the amount by which that spring is stretched. It is an example of a phenomenological law, a concise statement of an empirically observed regularity. In the 1930s, the theory of metals was worked out, and Hooke's law was explained in terms of the electromagnetic interaction between the atoms in a metal. Hooke's law addresses one specific phenomenon. In contrast, an understanding of fundamental laws governing electromagnetism enables us to explain a bewildering variety of phenomena.

When I was learning about such things as Hooke's law in high school, I got the impression that physicists try to find as many laws as possible, to explain every single phenomenon observed in the physical world. In fact, my colleagues and I in fundamental physics are working toward having as few laws as possible. The ambition of fundamental physics is to replace the multitude of phenomenological laws with a single fundamental law, so as to arrive at a unified description of Nature. This drive toward unity is *Fearful Symmetry*'s central theme.

2

Symmetry and Simplicity

I want to know how God created this world. I am not interested in this or
that phenomenon, in the spectrum of this or that element. I want to
know His thoughts, the rest are details.
—A. Einstein

A GLIMPSE OF NATURE

Suppose an architect wakes up to find himself imprisoned
in a strange room. He rushes to a window to look out. He can
glimpse a tower here, a column there: Evidently, he is in an enormous mansion. Soon, professional fascination overtakes the architect's fears. What he is able to see is beautiful. He is obsessed and
challenged; starting with what he has glimpsed, he wants to deduce
the underlying design of the mansion. Is the mansion's designer
a madman who piled complexity upon complexity? Did he stick a
wing here, a pediment there, without rhyme or reason? Is he a
hack architect? The architect-prisoner is sustained by an inexpli-
cable faith that the foremost architect in the world has designed
the mansion based on an elegantly simple and unifying principle.

We, too, wake to find ourselves in a strangely beautiful
universe. The sheer splendor and wealth of physical phenomena
never fail to astonish us. As physics progressed, physicists discovered that the diversity of phenomena did not require a diversity of
explanations. In this century physicists have become increasingly
ambitious. They have witnessed the incessant dance of the quantum and glimpsed the eternal secrets of space and time. No longer
content to explain this phenomenon or that, they have become
imbued with the faith that Nature has an underlying design of
beautiful simplicity. Since Einstein, this faith in the ultimate com-
prehensibility of the world has sustained them.

Fundamental physics progresses in spurts. Understanding
that has accumulated slowly is suddenly synthesized, and the en-

tire outlook of the field shifts. The invention of quantum physics in the 1920s furnishes a dramatic example. The years following 1971 will also likely come to be regarded as one of those spurts of feverish creativity from which deeper understanding emerges. In their exhilaration and unlimited hubris, some physicists have even gone so far as to suggest that we now have glimpsed the ultimate design of Nature, a claim that we will examine.

This glimpse reveals one astonishing fact: Nature's underlying design appears beautifully simple. Einstein was right.

AN AUSTERE BEAUTY

The term "beauty" is loaded with connotations. In everyday experience our perception of beauty is tied to the psychological, the cultural, the social, and, often even the biological. Evidently, that kind of beauty does not lie at the heart of physics.

The beauty that Nature has revealed to physicists in Her laws is a beauty of design, a beauty that recalls, to some extent, the beauty of classical architecture, with its emphasis on geometry and symmetry. The system of aesthetics used by physicists in judging Nature also draws its inspiration from the austere finality of geometry.

Picture a circle, a square, and a rectangle. Quick, which one is more pleasing to the eye? Following the ancient Greeks, most people will probably choose the circle. To be sure, the square, even the rectangle, is not without passionate admirers. But there is an objective criterion that ranks the three, circle, square, rectangle, in that order: The circle possesses more symmetry.

Perhaps I should not ask which geometrical figure is more beautiful, but which is more symmetrical. But again, following the ancient Greeks, who waxed eloquent on the perfect beauty of spheres and the celestial music they make, I will continue to equate symmetry with beauty.

The precise mathematical definition of symmetry involves the notion of invariance. A geometrical figure is said to be symmetric under certain operations if those operations leave it unchanged. For example, the circle is left invariant by rotations around its center. Considered as an abstract entity, the circle is unchanged, whether we rotate it through 17° or any other angle. The square, in contrast, is left unchanged only by rotations around its center through angles of 90°, 180°, 270°, and 360°. (As far as its effect on geometrical figures is concerned, a rotation through 360° is equivalent, of course, to a rotation through 0°, or no rotation at all.) The rectangle is even less symmetric than the square. It is left invariant only by rotations around its center through angles of 180° and 360°.

Besides rotations, reflections also leave these simple geometrical figures invariant. Once again, the circle is more symmetric; it is left invariant by reflections across any straight line that passes through its center.

There is an alternative, but equivalent, formulation of the notion of symmetry that is more convenient for physics. Instead of rotating a given geometrical figure, one can ask whether the figure appears the same to two observers whose viewpoints are rotated from each other. Obviously, if I tilt my head 17°, the square will look tilted but the circle will look the same.

THE BEAVER'S LESSON

You boil it in sawdust:
You salt it in glue:
You condense it with locusts in tape:
Still keeping one principal object in view—
To preserve its symmetrical shape.
—Lewis Carroll, "The Beaver's Lesson" in *The Hunting of the Snark*

In geometry, it is entirely natural to ask what one can do to a geometrical object without changing it. But physicists do not deal with geometrical figures. So how, then, does symmetry enter into physics?

Following the geometer, the physicist might want to ask what one can "do" to physical reality without changing it. This is clearly not quite the right question, but it suggests one of the basic concerns of physics: Does physical reality appear different as perceived by different physicists with different viewpoints?

Consider two physicists. Suppose one of them, for some nutty reason, always looks at the world with his head tilted 31° from the vertical, while the other takes the more conventional view. After years of study, the two separately summarize their observations in several physical laws. Finally, they compare notes.

Figure 2.1.   An artist's conception of the beaver's lesson.

We say that a physical law is invariant under rotation by 31° if they agree on that law. The nutty physicist now tilts his head to some other angle and resumes his study of the world. Eventually, the two physicists come to suspect that they agree regardless of the angle that separates their viewpoints. Real-life physicists have also come to believe that physical laws are invariant under rotation by any angle. Physics is said to have rotational symmetry.

ROTATIONAL SYMMETRY

Historically, physicists first became aware of the symmetries of rotation and reflection—the symmetries associated with the space we actually live in. In the next chapter, I will tell the strange story of reflection symmetry. Here, I will discuss rotational symmetry as a particularly simple and intuitively accessible example of a physical symmetry.

I have given a somewhat long, but precise definition of rotational symmetry: If we rotate our viewpoint, physical reality remains the same. The intellectual precision of our definition of rotational symmetry is necessary lest we make the same mistake as the ancient Greeks. I could have simply said that physical reality is perfect, like a circle or a sphere. Indeed, this vague but striking

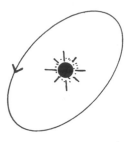 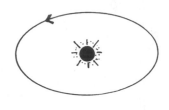

Figure 2.2.   In classical physics, rotational symmetry merely tells us that if we rotate a planetary system through any angle we choose, the rotated orbit is also a possible orbit. The ancient Greeks thought erroneously that rotational symmetry implies a circular orbit.

statement more or less paraphrases what the ancients believed, the sort of statement that lured them into the fallacious conclusion that the orbits of the planets must be circles. The correct definition of rotational symmetry does not require circular orbits at all.

Evidently, to say that physics has rotational symmetry is to say that it does not pick out a special direction in space. To the modern rational mind, particularly the mind entertained by films of intergalactic warfare, the statement that no one direction is intrinsically preferable to another is almost a matter of philosophical necessity. It seems absurd to point to some direction and say that *that* direction is special. But, in fact, not long ago everybody believed precisely that. For eons, human perception of the physical world was dominated totally by gravity, and the realization that the terms "up" and "down" had no intrinsic significance came as an astounding discovery. Though Eratosthenes in ancient Greece suspected that the earth was round, our understanding of rotational symmetry really started with Newton's insight that apples do not fall down to earth, but rather toward the center of the earth.

It goes without saying that physics is founded on empiricism and that rotational symmetry can only be established by experiment. In the 1930s, the Hungarian-American physicist Eugene Wigner worked out the observable consequences of rotational symmetry applied to quantum phenomena, such as the emission of light by an atom. No, experimentalists do not actually tilt their heads. They achieve the same effect, instead, by placing several light detectors around light-emitting atoms. The rates at which the various detectors register the arrival of light are monitored and compared with the theoretical rate Wigner predicted using rotational symmetry.

Thus far, experiments have always upheld rotational invariance. If tomorrow's newspaper were to report the fall of this cherished symmetry, physicists would be shocked out of their minds. At issue would be nothing less than our fundamental conception of space.

We intuitively know space to be a smooth continuum, an arena in which the fundamental particles move and interact. This assumption underpins our physical theories, and no experimental evidence has ever contradicted it. However, the possibility that space may not be smooth cannot be excluded. A piece of silver looks perfectly smooth and structureless to the eye, but on closer inspection, we see a latticework of atoms. Is space itself a lattice? Our experimental probes simply may not have been fine enough to detect any graininess to space itself.

Thus, physicists developed the notion of symmetry as an objective criterion in judging Nature's design. Given two theories, physicists feel that the more symmetrical one, generally, is the more beautiful. When the beholder is a physicist, beauty *means* symmetry.

SYMMETRY OF PHYSICAL LAWS

It is crucial to distinguish between the symmetry of physical laws and the symmetry imposed by a specific situation. For example, physics students, traditionally, are made to work out the propagation of electromagnetic waves down a cylindrical metal pipe. While the laws of electromagnetism possess rotational symmetry, the problem obviously has only cylindrical symmetry: The axis of the pipe defines a direction in space. Physicists studying specific phenomena generally are more aware of the symmetry imposed by physical situations than of the intrinsic symmetry of the physical laws. In contrast, in this book we are interested in the symmetry of fundamental laws. To underscore the point, let me give another example. In watching an apple fall, the fundamental physicist is interested in the fact that the law of gravity does not pick out any special direction, rather the fact that the earth is nearly spherical. The earth could be shaped like an eggplant, for all the physicist cares.

This distinction between the symmetry of physical laws and the symmetry imposed by a specific situation was one of Newton's great intellectual achievements, and it enabled physics as we know it to take shape. While this distinction, once spelled out, is fairly obvious, it is easy to get confused since in everyday usage we invariably mean by symmetry the symmetry of specific situations. When we say a painting displays a certain symmetry, we refer to the symmetry in the artist's arrangement of the pigments, which has nothing to do, of course, with the symmetry of the physical laws governing the pigment molecules. In this book, I try to explain abstract concepts by analogies involving concrete objects. The reader must keep in mind that we are always interested in the symmetry of the physical laws rather than the symmetry of concrete objects.

SPRING REDUX

In introducing this chapter, I said that physicists have glimpsed both beauty and simplicity in Nature's design. What do physicists mean by simplicity?

In its drive toward simplicity, the march of physics has been relentlessly reductionistic. Physics is possible because complicated phenomena can be reduced to their essentials.

Historically, for physics to progress, many *why* questions had to be reexpressed as *hows*. "Why does a stone accelerate as it falls?" The ancients thought that the stone, like a horse, is eager to return home. Physics began when Galileo, instead of asking why the stone fell, went out and measured how.

As children we are full of whys. But every answered why is replaced by another why. "Why are leaves so pleasantly green in the spring?" Well, the professor explains, leaves contain the chlorophyll molecule, a complicated assemblage of atoms that interacts with light waves in a complex way. The chlorophyll molecule absorbs most of the light, but not those components the human eye perceives as green. The explanation bores the typical layman (and nowadays, many physicists as well). Ultimately, the explanation to this question, and to numerous others like it, boils down to how the electron interacts with the fundamental particle of light, the photon.

Physicists began the modern theory of the interaction between electron and photon around 1928 and completed it by the early 1950s. How the electron interacts with the photon has been thoroughly understood for more than thirty years, and yet, one can't help but wonder why these two fundamental particles interact in the rather peculiar fashion that they do. This question, too, has been answered. Physicists know now that the electron-photon interaction is completely fixed by a symmetry principle, known as the gauge principle, which plays a pervasive role in Nature. Evidently, physicists can now insist on asking why Nature should respect the gauge principle. Here contemporary physics stops, and any discussion on this question, which amounts essentially to asking why there is light, dissolves into a haze of speculation.

While whys have been replaced by other whys, enormous progress has been made: One why replaces many whys. The theory of electron-photon interaction enables us to explain not only the verdure of spring, but also the stretching of springs, not to mention the behavior of lasers and transistors. In fact, almost all the physical phenomena of which we are directly aware may be explained by the interaction between photons and electrons.

Physics is the most reductionistic of sciences. In contrast, the explanations that I have read in popular expositions of biology, although fascinating, have been emphatically nonreductionistic. Often, the explanations in terms of biochemical processes are more complicated than the phenomena in question.

Contemporary physics rests on the cornerstone of reductionism. As we delve deeper, Nature appears ever simpler. That this is so is, in fact, astonishing. We have no a priori reason to expect the universe, with its fantastic wealth of bewilderingly complex phenomena, to be governed ultimately by a few simple rules.

SIMPLICITY BEGETS COMPLEXITY

Let the next contest for the Prix de Rome for architects ask for a design of the universe. Asked to design the universe, many would be tempted to overdesign, to make things complicated so that their universe would display an interesting variety of phenomena.

It is easy to produce complicated behavior with a complicated design. As children, when we take apart a complicated mechanical toy, we expect to find a maze of cogs and wheels hidden

Figure 2.3.   Football has complicated rules, the game of Go exceedingly simple ones.

inside. The American game of football is my favorite sport to watch, because of the variety of behavior exhibited. But the complex repertoire is the direct result of probably the most complicated set of rules in sports. Similarly, the complexity of chess is generated by its rather complicated rules.

Nature, whose complexity emerges from simplicity, is cleverer. One might say that the workings of the universe are more like the oriental game of Go than chess or football. The rules of Go can be stated simply and yet give rise to complex patterns. The eminent physicist Shelley Glashow has likened contemporary physicists to kibitzers at a game whose rules they do not know. But by watching long and hard, the kibitzers begin to guess what the rules might be.

As glimpsed by physicists, Nature's rules are simple, but also intricate: Different rules are subtly related to each other. The intricate relations between the rules produce interesting effects in many physical situations.

In the United States, a committee of the National Football League meets every year to review the past season and to tinker with the rules. As every observer of the sport knows, an apparently insignificant change in even one rule can drastically affect the pattern of the game. Restrict slightly how a cornerback can bump a receiver and the game becomes offense dominated. Over the years, the rules of the game have evolved to ensure an interesting balance between offense and defense. Similarly, Nature's laws appear to be delicately balanced.

An example of this balance occurs in the evolution of stars.

A typical star starts out as a gas of protons and electrons. Under the effect of gravity, the gas eventually condenses into a spherical blob in which nuclear and electric forces stage a mighty contest. The reader might recall that the electric force is such that like charges repel each other. Protons are kept apart, therefore, by their mutual electric repulsion. On the other hand, the nuclear attraction between protons tries to bring them together. In this struggle the electric force has a slight edge, a fact of great importance to us. Were the nuclear attraction between protons a tiny bit stronger, two protons could get stuck together, thus releasing energy. Nuclear reactions would then occur very rapidly, burning out the nuclear fuel of stars in a short time, thereby making steady stellar evolution, let alone civilization, impossible. In fact, the nuclear force is barely strong enough to glue a proton and a neutron together, but not strong enough to glue two protons together. Roughly speaking, before a proton can interact with another proton, it first has to transform itself into a neutron. This transformation is effected by what is known as the weak interaction. Processes effected by the weak interaction occur very slowly, as the term "weak" suggests. As a result, nuclear burning in a typical star like the sun occurs at a stately pace, bathing us in a steady, warm glow.

The central point is that, unlike the rules of football, Nature's rules are not arbitrary; they are dictated by the same general principle of symmetry and linked together in an organic whole.

Nature's design is not only simple, but minimally so, in the sense that were the design any simpler, the universe would be a much duller place. Theoretical physicists sometimes amuse themselves by imagining what the universe would be like if the design were less symmetric. These mental exercises show that not a stone can be disturbed lest the entire edifice crumble. For instance, light might disappear from the universe, and that would be no fun at all.

THE RULE OF LARGE NUMBERS

One reason that an underlying simplicity can generate complicated phenomena is the occurrence in Nature of ridiculously large numbers. A drop of water contains a mind-boggling number of atoms. Young children are fascinated by large numbers and are delighted when taught words like "thousand" and "million." They want to know if there are numbers larger than a million. My three-year-old son was pleased to learn that infinity is the largest number. But to young children, the words "thousand," "million," and even "one hundred" are synonymous with "many." I am reminded that George Gamow, the great Russian-American physicist who first suggested that the universe started with a big bang and who splendidly popularized physics, once told of a Hungarian count whose counting range was limited to one, two, three, and many.

While physicists can talk of and compute large numbers, the human mind staggers and cannot truly grasp the reality associated with the gigantic numbers Nature plays with. I cannot even comprehend the relatively small large numbers I read about in the newspapers unless I reduce them to per capita figures. By carrying out this exercise, the reader will be delightfully surprised to discover how often the figures cited in the popular press turn out to be nonsensical.

A sociologist is certainly not surprised that systems with a large number of particles can behave quite differently from systems with few particles. In this electronic age, we oblige electrons to rush about in a controlled collective frenzy. To record one beat of rock video on tape, more atoms than the number of people on earth have to line up in just the right formation.

As children, we wondered why so many grains of sand stretched along the beach. Some of the deepest thinkers in physics have also wondered: Why is the universe populated by so many particles?

The issue of the number of particles in the universe is logically quite distinct from the issue of simplicity of design. After devising some simple laws governing the interaction between particles, a designer of the universe in the imagined Prix de Rome contest may decide to throw in a reasonable number of particles, say three protons and three electrons. Perhaps he or she would also throw in a few particles of light, say seven photons. Such a universe, of course, would be rather uneventful, but this possibility is not excluded, logically. Instead, the number of protons within the observable universe is estimated at about 10^{78}, and the number of photons, 10^{88}. The reader probably knows that the number 10^{78} can also be written as the number 1 followed by seventy-eight zeroes. These numbers are absurdly large! Who ordered this many protons, anyway?

So, the universe contains a zillion particles. Why? This question, sometimes referred to as the "population question," is intimately related to the "vastness question" and the "longevity question." Why is the universe so large, and so old? The universe has lived a long time when measured by the duration over which subnuclear particles live and die. Why doesn't the universe expand and collapse, all within this fundamental time scale exhibited in Nature's laws? Until recently, most physicists regarded these questions as unanswerable. But thanks to exciting recent developments to be described in later chapters, some physicists think they may have some of the answers.

A HIERARCHY OF STRENGTHS

Mysteriously enough, not only is there a large number of particles in the visible universe, but the basic laws themselves display large numbers. According to modern physics, there are four fundamental interactions between particles: the electromagnetic, the gravitational, the strong, and the weak.

The electromagnetic interaction holds atoms together, governs the propagation of light and radio waves, causes chemical reactions, and prevents us from walking through walls and sinking through the floor. In an atom, electrons, with their negative electric charges, are prevented from flying off because of their attraction to the positive charges carried by protons located in the nucleus. The gravitational interaction keeps us from flying off into space, holds planetary systems and galaxies together, and controls the expansion of the universe. The strong interaction holds the nucleus of the atom together, the weak causes certain radioactive nuclei to disintegrate. Although of fundamental importance in Nature's design, the strong and weak interactions do not appear to play a role in any phenomenon at the human scale. As we saw earlier, all four interactions play crucial roles in stellar burning.

As the names strong and weak suggest, one interaction is considerably stronger than the electromagnetic interaction, the other considerably weaker. But most dramatically, the gravitational force is far, far weaker than the other three. The electric force between two protons is stronger than the gravitational attraction by the enormous ratio of 1 to about 10^{38}, another absurdly large number.

Ironically, we are normally most aware of gravity, by far the most feeble force in Nature. Although the gravitational attraction between any two atoms is fantastically small, every atom in our bodies is attracted to every atom in the earth, and the force adds up. In this example, the incredibly large number of particles involved compensates for the incredible weakness of gravity. In contrast, the electric force between two particles is attractive or repulsive according to the signs of the electric charges involved. A lump of everyday matter contains almost exactly an equal number of electrons and protons, so the electric force between two such lumps almost cancels out.

Physicists refer to this peculiar arrangement of these four enormously different interactions as the hierarchy of interactions. Incidentally, "hierarchy" originally referred to the system invented by Dionysius the Areopagite to arrange angels into three divisions, each consisting of three orders. Nowadays, we arrange the fundamental interactions of matter instead.

Nature was very considerate to arrange Herself hierarchically. When studying one interaction, physicists can usually neglect the others. In this way, they are able to disentangle the four. Because reality is arranged in layers, like an onion, we can learn about Nature in increments. We can understand the atom without understanding the atomic nucleus. Atomic physicists do not have to wait for nuclear physicists, and nuclear physicists do not have to wait for particle physicists. Physical reality does not have to be understood all at once. Thank you, Nature.

The two conceptually distinct occurrences of large numbers in Nature, the large population of particles and the large disparity in the strengths of fundamental interactions, create the awesome range of scales in the universe, from the unspeakable void between galaxies, across which it takes light eons to travel, to the barely conceivable distance between the atoms in a drop of water. We humans occupy the middle ground, between the microscopically small and the cosmically large, between the ephemeral and the almost eternal. In one second, as many lifetimes of certain unstable elementary particles have elapsed as seconds since the birth of the universe. We dwarf atoms as galaxies dwarf us. Within the time scale of human experience, ranging from about a tenth of a second to a hundred years, we live and die and create our art and science.

GUIDED BY SYMMETRY

We can now see how Nature's underlying design can be simple and comprehensible in spite of the apparent complexity of the physical world. Later, I describe how physicists are beginning to decipher this underlying design by postulating various symmetries that Nature may have used in Her design.

Historically, physicists first became aware of those symmetries associated with the space in which we actually live; namely rotation and reflection. The next chapter tells the strange story of reflection symmetry. Because rotation and reflection are firmly rooted in our intuitive perception, our discussion of rotational symmetry is not expressed in the full power and glory of the physicist's language of symmetry. Further on, when we encounter the more abstract symmetries in Nature's design, this symmetry language becomes indispensable.

3

The Far Side of the Mirror

Dear Miss Manners:
  Which way does one pass the food, to the right or left?

Gentle Reader:
  Food platters should travel left to right.

MIND AND SYMMETRY

In the last few years, I have often watched my son Andrew and his friends play with blocks. Children up to a certain age simply pile block on top of block, but then, in one of those developmental leaps described by Piaget, they suddenly begin to erect structures that display a pronounced left-right symmetry. Those children who grow up to become architects eventually end up building structures like those illustrated in Figures 3.1, 3.2, and 3.3. Architecture is practically founded on the tenet of bilateral symmetry. Asymmetrical buildings are regarded as oddities and demand explanation. For example, the cathedral at Chartres is amusingly asymmetrical. Its construction took so long that architectural styles changed.

Not surprisingly, modern architecture, in tune with the rebellious character of our century, has spawned a number of dramatically asymmetrical buildings, but postmodernism, the architectural movement now in vogue, is partly a movement to revive certain classical principles, such as bilateral symmetry.

Given that the human body has such a pronounced bilateral symmetry, the notion of the world being divided into a left side and a right must come to us in early childhood. Obviously, biological evolution imposed left-right symmetry on the human body and on most animal forms. The symmetric arrangement of ears and eyes clearly is necessary for stereo reception, of legs, for locomotion in a straight line. Even the interstellar denizens we see in

Figure 3.1.   The striking symmetry of the Taj Mahal contributes to its architectural splendor. *(Lauros-Giraudon, courtesy Art Resource, NY)*

Figure 3.2.   The Portland Public Services Building in Portland, Oregon, designed by Michael Graves, exemplifies the reemergence of symmetry in postmodern architecture.   *(Michael Graves: Buildings and Projects,*   Published by
Rizzoli, New York)

View from Fifth Avenue

Figure 3.3.   Chartres Cathedral: The Romanesque spire *(on the right)* was built in the twelfth century, the Gothic spire *(on the left)* in the sixteenth century. It has been said that the two asymmetric spires at Chartres marked the beginning and end of architecture in the Middle Ages. *(Lauros-Giraudon, courtesy Art Resources, NY)*

movies, interestingly, tend to be endowed with a left-right symmetry. Indeed, left-right symmetry is so prevalent in the biological world that we find any deviations from it odd and fascinating.

The division of the human brain into left and right halves, each with different functions, is well known. That one of the ovaries in a chicken is atrophied and nonfunctional provides another example. Perhaps the most astonishing example I know of involves the *Poeciliidae* family of small fish that live in tropical American waters. I quote from a description given by Guy Murchie:

> Their most unusual feature is the male sex organ, which evidently evolved from a ventral fin and can be half as long as its owner. In erection it enlarges and swings forward until, in some species, its tip is almost even with the fish's nose yet pointing perhaps 30° to the right or left. In several species this fishy phallus has fingerlike appendages that one can imagine must be delightfully handy for feeling its way into the female and it is sometimes also abetted by two sets of comblike retorse spines (apparently evolved from side-fins) for clasping her the while. But she definitely must have her orifice on the correct side, right or left, to receive the male, else the whole match is off.

The human mind finds pleasing the economy of design associated with bilateral symmetry. We simply have to look around at any number of common objects to see how often designers adhere to this principle. But the human mind is also capable of making rather strange associations.

The annals of Western painting abundantly illustrate these two bents. Look at a typical Renaissance religious painting, with its rigidly symmetrical placement of a saintly pair on either side of the central sacred subject. Normally, the saint to the right of the subject is higher in the sainthood pecking order than the one on the left. In a case where the patrons, often man and wife, are also portrayed, the male, almost invariably, is kneeling on the right. Another convention is that the light in a classical painting usually comes from the right of the subject. Interestingly enough, many well-known artists, when reaching for the mass market, quite willingly allow their art to violate these conventions. Rembrandt, for example, did not bother to make the necessary adjustments so that his etchings would come out obeying the standard convention of right-over-left. At this point, I would like to give the reader a quiz. Form in your mind's eye the well-known image of Michelangelo's portrayal of the creation of man on the Sistine Chapel ceiling. Did God touch Adam with His right hand or His left?

On a man's jacket, the buttons are on the right, while on a woman's they are on the left. The standard explanation is that when caught in a bind, a man could rip open his jacket with his left hand and with a sweeping motion, draw his sword with his right. It would also be easier for a right-hander to manipulate buttons sewn on the right-hand side. However, a lady was dressed and undressed, not by herself of course, but by someone else, her lady-in-waiting, perhaps; hence, a left-hand approach was easier.

ALICE AND NARCISSUS

Let us now turn to physics. Does Nature, like dinner guests of old, care about the difference between left and right? If Nature does not care, then physicists say that the world is parity invariant or reflection invariant. Let me be precise here and provide an operational definition of parity invariance. Take your favorite physical phenomenon—anything from two billiard balls colliding to an atom emitting light. Put a mirror in front of what is happening and ask, Does the process we observe in the mirror contradict the laws of Nature as we know them? If not, we say that the laws governing that process are parity invariant. This definition is carefully worded to exclude any reference to left-right asymmetry that is not of intrinsic physical interest.

To say that physics is parity invariant is *not* to say that the world inside the mirror is the same as our world. When I look at myself in the mirror, I see a person who looks like me. But his heart is on his right side, his watch hands sweep counterclockwise. Even the double helixes of his DNA molecules coil in the other direction. The point is that the laws of physics do not forbid the existence of a person with his heart on his right. If we had always fed him (and his ancestors) biological molecules that are mirror images of ours, his double helixes really would coil the other way. While it is beyond the biologist's ability to construct such an individual, the clockmaker could easily construct a watch whose hands would sweep counterclockwise. It would be governed by the parity invariant laws of physics and would keep precisely the same time.

To physicists, the fact that our heart lies slightly on our left is of no intrinsic significance, merely an accident of biological evolution. Some early clockmakers had simply agreed on the convention that hands on clocks sweep, well, clockwise. Similarly, that certain organic molecules spiral one way or the other is regarded to be of no fundamental significance. Chemists are able to construct mirror images of molecules found in Nature, and these molecules, indeed, have the same physical properties. One can easily imagine that at the dawn of life both types of organic molecules were present. Because of statistical fluctuation, one type happened to be slightly more numerous than the other and ended up dominating and driving the other type into extinction.

In *Through the Looking Glass,* the sequel to *Alice in Wonderland,* Lewis Carroll articulated a fantasy that most of us have had, particularly as children. I have observed with great interest the relationship my young son had with his mirror image. At some point, a young child rather suddenly appears to realize that the mirror image is not an independent person, and thereafter looks into the mirror with the eyes of an adult. Narcissus is evidently a strange character.

Alice climbed through the mirror on the mantel and found herself in another world. Carroll's fantasy offers us a sharp way of stating the notion of parity invariance. Let us follow Alice into the mirror world.

Everything looks slightly and amusingly different, but that does not concern us. We want to find a physicist to ask him what he knows of the fundamental interaction between particles. If his version of the physical laws agrees with ours, we would conclude that Nature does not distinguish between left and right.

A CHERISHED BELIEF SHAKEN

If I were to poll the man on the street as to whether the basic design of Nature is left-right symmetric, I suspect that I would receive, besides the "no opinion" and "who cares" that afflict any pollster's life, at least a few "perhaps not." Yet, until the year 1956, physicists held it as self-evident that Nature does not distinguish between left and right. Physicists of the nineteenth century had subjected this belief to experimental tests and had found no sign that Nature would favor left over right, or vice versa.

With the advent of atomic and nuclear physics in the early decades of the twentieth century, the assumption of parity invariance again was tested in a number of experiments. By the mid-1950s, parity invariance was regarded universally as one of the few sacred principles dear to physicists, who were loath to think that Nature would favor right over left or vice versa. The idea that Nature would subscribe to the same kind of arbitrary convention as a society hostess seating an honored guest on the right seemed absurd. But the physics community was soon to be shocked.

By the mid-1950s, physicists had discovered a number of new particles whose existence was so unexpected that they were dubbed "strange particles" in exasperation. Using the newly built accelerator at the Brookhaven National Laboratory, Long Island, New York, experimentalists studied the strange particles with great care. By late 1955, thanks largely to detailed analysis by R. H. Dalitz, an Australian physicist, it became clear that some of the strange particles were exhibiting very puzzling behavior in the way they decayed.

In April 1956, at a conference on high energy physics held at Rochester, New York, the puzzle of the strange particles was much discussed, but none of the proposed resolutions of the puzzle was satisfactory. The Chinese-American physicist C. N. Yang gave a summary lecture on the strange particles. After Yang's lecture, a spirited discussion ensued. At that point, Richard Feynman brought up a question that Martin Block had asked him, whether parity invariance, which Dalitz used as an implicit assumption in his analysis, should be questioned. Yang replied that he and another Chinese-American physicist, T. D. Lee, had been analyzing that possibility but had not yet reached a conclusion.

With hindsight, it is easy to see that parity violation—the notion that Nature would distinguish between right and left— would offer a natural way out of this puzzling situation. However, the idea of a left-right symmetric Nature was so deeply ingrained that parity violation was considered the least likely answer to the mystery.

Lee and Yang continued to struggle with the problem. Yang recalled later that he felt like "a man in a dark room groping for an outlet." In early May 1956, Yang came to visit Lee, and the two of them ended up driving around Columbia University, where Lee was a professor, in an unsuccessful attempt to find a parking space. While they were going round and round, Lee and Yang discussed the possibility of parity violation. Finally, in exasperation, they gave up and double-parked in front of a Chinese restaurant. The double frustration of wrestling with the mystery of the strange particles and of simultaneously looking for a parking space must have done peculiar things to their minds, for history records that when they sat down, they were struck by the crucial point that the experimental evidence in support of parity invariance comes from processes that involve either the electromagnetic force, such as in the emission of light by an atom, or the strong force, such as in the collision of two atomic nuclei. The decay of strange particles, in contrast, had been determined by 1956 to be governed by weak force, known to be responsible for certain radioactive decays of atomic nuclei.

The crucial idea which came to Lee and Yang is that Nature may respect parity in many of Her laws, but not in the laws governing the weak interaction between particles. Imagine that one of the fundamental principles of our legal system, that the accused is presumed innocent until proven otherwise, is decreed to hold only for certain crimes, while for other crimes the opposite is to be the case. Just as judicial philosophers would surely cringe at this notion, physicists find Nature's selective violation of parity rather discomfiting philosophically.

For the next few weeks, Lee and Yang engaged themselves in a detailed analysis of the numerous completed experiments involving the weak interaction and concluded that a possible parity violation would not have shown up in any of these experiments. Their next task was to invent an experiment sensitive to parity violation. In June, they published their now-historic paper, questioning parity invariance in the weak interaction and outlining experiments that would settle the question.

IS THE MIRROR WORLD THE SAME AS OURS?

One of the experiments Lee and Yang proposed involved a spinning nucleus.

Many species of atomic nuclei are permanently spinning inside the atom. As the reader knows, the atom resembles a miniature solar system. The nucleus is the sun, around which electrons orbit like planets. The radii of the electron orbits are so much larger than the size of the nucleus that the electrons play no role in our subsequent discussion. They are far away.

Before proceeding further, I have to explain how physicists define a direction of spin for any spinning object. Imagine curling your left hand around a given spinning object so that the fingers are pointing along the direction in which the surface of the spinning object is revolving. We define the direction of spin as the direction the thumb is pointing. For instance, in Figure 3.4 A the ballerina is said by a physicist to be spinning "up," in Figure B, spinning "down." ("Up" and "down," in this example, are defined with respect to the surface of the earth, of course, but the definition clearly works even for an object spinning deep in space.) The use of the left hand is purely a matter of convention, akin to the fact that, in some countries, one drives on the right-hand side of the road, while in others one drives on the left. The importance here is to have a convenient way of specifying which way the object is spinning. One could have just as well spoken of clockwise and counterclockwise, but that depends on which end of the spinning object we look at. Let me illustrate this principle by that wonderful imagery from American football, the quarterback throwing a long pass. The quarterback and the receiver will disagree as to whether the ball is spinning clockwise or counterclockwise.

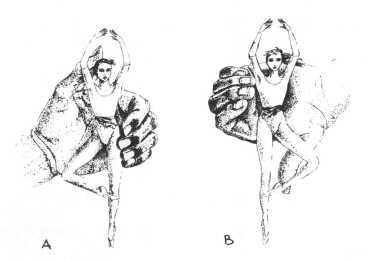

Figure 3.4. Note how the artist used visual cues to indicate the directions in which the two ballerinas are pirouetting. According to the convention described in the text, the ballerina on the left (A) is spinning up, while the ballerina on the right (B) is spinning down.

The reader should understand that there is absolutely nothing profound about the preceding. The long discussion is necessary so that we will know exactly what we are talking about next.

Lee and Yang suggested studying the decay of a spinning radioactive nucleus. A nucleus may be thought of as a collection of protons and neutrons stuck together. In a radioactive nucleus the arrangement of protons and neutrons is not quite stable, and, in a given period of time, there is a certain probability that the radioactive nucleus will decay. If the weak force is responsible for the decay, then the probability per unit time is extremely small. That is precisely why the force is called weak. The decaying nucleus shoots off an electron and another particle, a neutrino, which will not be detected in the experiment. The electron flies off at high speed. This electron, emitted with the decay, is not to be confused with the electrons that have been orbiting the nucleus some large distance away.

As previously explained, the spinning nucleus establishes a direction. We can now ask, Does the electron come shooting out in this direction or in the opposite direction? To see how the answer to this question will reveal whether Nature violates parity, we apply the criterion explained earlier and compare what happens in our world with what would happen in the mirror world.

Suppose the electron emerges in the direction of the nuclear spin. Now, look inside the mirror. (See Figure 3.5.) Just as the hands on the watch in the mirror sweep counterclockwise, the nucleus in the mirror is spinning in the opposite direction. In the mirror world, the electron emerges in the direction opposite to the nuclear spin! A physicist observing this decay, and his colleague, likewise observing in the mirror world, will reach totally opposite conclusions about the law governing how the electron emerges in a radioactive decay. If Nature respects parity, then electrons should emerge with equal probability in the direction of the nuclear spin and in the opposite direction. In the actual experiment, many nuclei are involved, and one observes the electrons emerging from many decays to see if they prefer to emerge in one direction or the other.

It is clearly necessary that the nucleus must spin in order to establish a reference direction. (This does not mean, however, that parity violation can only be observed in a process involving spinning particles.) It is also worth emphasizing that this proposed experiment does not involve strange particles in any way, so its

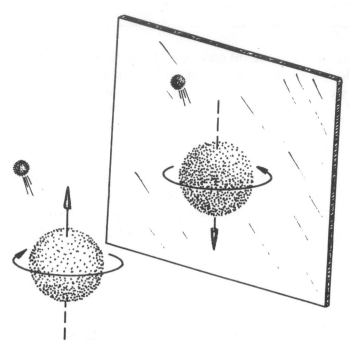

Figure 3.5. A spinning nucleus (represented as a large ball) ejects an electron (the small ball). In our world the electron emerges more or less in the direction of the nuclear spin; in the mirror world it emerges in a direction opposite to that of the nuclear spin. In the actual experiment the direction of the ejected electron relative to the nuclear-spin direction was tabulated statistically for a large number of nuclei. If the electron emerges preferentially in the nuclear-spin direction (as is suggested by the figure), then we can conclude that Nature violates parity invariance because a physicist in the mirror world would see the electron emerging preferentially in a direction opposite to the nuclear-spin direction. Our world and the mirror world would then be governed by different physical laws.

interpretation is not clouded by the then-unknown dynamics of strange particles.

THE LADY AND THE LEFT HAND OF GOD

The next step for Lee and Yang was to persuade someone capable of doing the experiment to do it. Physics journals are full of proposed experiments, but an experimentalist has to be convinced that the significance of the experiment warrants the enormous amount of effort required.

It was painless for Ptolemy to speculate that the source of the Nile lies in the heart of Africa, but it was another matter for Burton and Speke to devote their lives and sanities to the question.

After approaching a number of experimentalists, many of whom were skeptical, Lee and Yang managed to talk Chien-Shiung Wu, one of the leading authorities on weak interaction experiments, into taking a chance on a long shot.

Madame Wu, as she is universally known in the physics community, is a remarkable figure. Born in China in 1912, only one year after the demise of the infamous Manchu dynasty, she came to be known as the "reigning queen of experimental nuclear physics," and she served as the first woman president of the American Physical Society, blazing trails for women experimenters in a male-dominated field. Her experiments are characterized by a meticulous care and a stylish simplicity that some of her colleagues have described as feminine. Madame Wu was so intrigued by what Lee and Yang had to say that she canceled her summer trip and started work immediately. Thus it came to pass that Nature first revealed her "handedness" to a lady.

Madame Wu, like Alice before her, looked inside the mirror. In so doing, she met with a number of obstacles. Although things look quite simple to theorists (see Figure 3.5), the real-life complications that experimentalists must address are quite formidable. For instance, no one handed Madame Wu a single spinning nucleus. The enormous number of nuclei contained in the experimental sample are all spinning in different directions. The experiment would work only if she could, somehow, line up all the nuclear spins in the same direction. At room temperature, atoms vibrate about in perpetual agitation, so that the nuclear spins, once lined up, would soon again point in different directions. Hence, she had to conduct the experiment at extremely low temperatures to minimize the thermal agitation of the atoms. Sophisticated refrigeration devices, pumps, and so forth, all had to be brought in, and we all know how such machines are apt to malfunction. (Experimental and theoretical physics attract rather different personality types, with different temperaments and abilities. Sociologists should find here a ripe field for a fruitful study.) Madame Wu, therefore, contacted and collaborated with a team of low-temperature physicists at the National Bureau of Standards in Washington, where the required refrigeration device was available.

By December 1956, she and her coworkers had found strong evidence that parity is in fact violated. In decays governed by the weak force, electrons prefer to emerge in one direction over another. Independently, a group led by Valentine Telegdi at the University of Chicago had arrived at the same conclusion by doing another experiment proposed by Lee and Yang.

On Friday, January 4, 1957, Lee described the definitive result of Wu's experiment to a group of his colleagues. The discussion was particularly animated during lunch when Leon Lederman, an experimentalist at Columbia, suddenly realized that, with luck, he might be able to detect parity violation in the decay of the pi meson, a subnuclear particle then known for a number of years. Later that evening, he called Richard Garwin, a noted experimentalist now at IBM. By two in the morning, the two excited physicists had designed and set up the experiment, and proceeded to take data. But just when they thought they too had seen the left hand of God, the equipment broke down. They enlisted the help of another experimentalist and together they repaired the equipment, then set to working around the clock. At six o'clock on Tuesday morning, Lederman called T. D. to say that Nature is unmistakably handed.

Modern particle experiments are normally mammoth multinational efforts involving, sometimes, a hundred or more physicists and lasting for years. The experiment of Lederman et al. is surely the shortest on record. Leon Lederman is now the director of the gigantic Fermi National Accelerator Laboratory in Batavia, Illinois. One imagines that he gets things done.

The news of parity violation stunned the physics community. It was as if *la plus grande dame* of etiquette had committed an unspeakable faux pas. The public was fascinated. For instance, Ben-Gurion, then prime minister of Israel, asked Madame Wu how parity relates to yoga. *The New York Times* editorialized on the meaning of parity. The news filtered slowly through society, becoming totally garbled and misunderstood. When I was a boy, a businessman friend of my father's told me that two Chinese-American physicists had overthrown Einstein's theory of relativity, whatever that was.

THE CURMUDGEON AND HIS GHOST

The discovery of parity violation profoundly altered our preconceived notions of Nature. But it also had an immediate and far-reaching impact on our understanding of the physical world. It turns out that parity violation was the missing ingredient needed for constructing a theory of weak interaction.

To understand the state of weak interaction theory in 1956, we have to go back to the early 1930s, when the English physicist C. D. Ellis carefully measured the speed of the electron ejected in the decay of a radioactive nucleus. This involves the same physical process examined by Madame Wu and company, but as is often the case in physics, different physical quantities are measured in different experiments. Ellis did not have the difficult task of lining up the radioactive nuclei; on the other hand, he had to measure the energy of the electron accurately, which Madame Wu did not have to do.

Ellis became a physicist under unusual circumstances. An army officer in World War I, he was captured early on. In the prison camp, he befriended a luckless fellow Englishman, James Chadwick. Young Chadwick, whom we will meet again in a starring role in a later chapter, had gone to Berlin to study radioactivity under Fritz Geiger, of the counter fame. When war broke out, the Germans arrested him as a spy. Out of sheer boredom, Chadwick proceeded to teach Ellis physics. Ellis was so fascinated that he abandoned his military career after the war.

When Ellis did his experiment, theorists thought they knew what energy the ejected electron should have. After all, the famous Albert Einstein told us how mass could be converted into energy according to his formula $E = mc^2$. Knowing the mass of the radioactive nucleus, and the mass of the daughter nucleus, one can determine, using simple subtraction and Einstein's formula, the energy that the electron should come out with. Call it E^*.

Surprise! Ellis found that the electron does not always come out with the same energy (although its energy is always less than E^*). In one decay, the electron might come out slowly, in another, much faster. Rarely would it have the energy E^*. Where did the missing energy go? Could Einstein possibly be wrong?

The resolution of this conundrum was given by Wolfgang Pauli, the jovial and rotund Viennese physicist who played the self-appointed role of curmudgeon in the drama of twentieth-century physics. Pauli was the master of the devastating put-down. It was said that Pauli, when told of a new theoretical result, would remark sadly, "It is not even wrong." He also went around lamenting that physics is too difficult, that he should have been a comedian instead. Of the many stories about Pauli circulating in the physics community, a favorite tells of Pauli, after his death, asking God to reveal His design (a standard fantasy among physicists). When God told him, Pauli exclaimed, "It is not even wrong."

In 1933, Pauli suggested that a hitherto unknown particle, which interacts neither strongly nor electromagnetically and thus would escape detection, carries away the missing energy, like a black-clad thief disappearing into the night. The mysterious particle, later to be given the Italian name "neutrino," was the first particle predicted to exist before it was actually discovered. Nowadays, in an age when particle theorists hypothesize the existence of experimentally unknown particles with wanton abandon, Pauli's boldness can only be appreciated in its historical context.

Pauli deduced that the neutrino has fantastic properties. In quantum mechanics, one speaks of probabilities. Since the neutrino is postulated to interact only via the weak force, the probability is very small that it interacts with an electron or a nucleus it encounters. (That is precisely why the weak force is called weak.) Knowing how weak the weak interaction is, Pauli concluded that a neutrino, like a ghost, can pass through the entire earth without interacting. On the other hand, we who are of flesh and blood cannot walk through walls because the probability that the atoms in our bodies would interact electromagnetically with atoms in the walls borders on certainty.

Ever critical of himself as well as of others, Pauli wrote to a friend that he had committed the worst mistake a physicist could commit: postulating a particle that cannot be subjected to experimental scrutiny. But he was overly pessimistic. In 1955, American physicists F. Reines and C. Cowan managed to "see" a neutrino. Nowadays, particle accelerators routinely shoot out beams of neutrinos and a few of these have been observed interacting with other matter. (To produce a neutrino beam, experimentalists first produce a beam of subnuclear particles that decay in flight into neutrinos.) The reader may well wonder how that is possible. The probability that a neutrino would interact with a nucleus, while almost inconceivably small, is not zero. To beat the small probability, one can pile an enormous number of nuclei in front of a beam of neutrinos, and wait. Once, the U.S. Navy junked some old battleships and gave the scrap iron to experimentalists. Even with that huge pile of iron, the experimentalists had to wait for months before they caught one neutrino interacting with an atom.

Pauli also deduced that the neutrino is massless because the electron in Ellis's experiment does have the energy E^* once in a while. If the neutrino had mass, then according to Einstein, part of the available energy E^* must be budgeted to produce the neutrino mass, thereby leaving a smaller amount for the electron. Knowing how the electron, the radioactive nucleus, and the daughter nucleus (which the radioactive nucleus decays into) spin, Pauli also concluded that the neutrino is endowed with a perpetual spin. The American novelist John Updike was so fascinated by the neutrino that he wrote a poem about it, to the best of my knowledge the only poem ever written about a subatomic particle by a major literary figure.

> Neutrinos, they are very small.
>   They have no charge and have no mass
> And do not interact at all.
>   The earth is just a silly ball
> To them, through which they simply pass,
>   Like dustmaids down a drafty hall
> Or photons through a sheet of glass.
>   They snub the most exquisite gas,
> Ignore the most substantial wall,
>   Cold-shoulder steel and sounding brass,
> Insult the stallion in his stall,
>   And scorning barriers of class,
> Infiltrate you and me! Like tall
>   And painless guillotines, they fall
> Down through our heads into the grass.
>   At night, they enter at Nepal
> And pierce the lover and his lass
>   From underneath the bed—you call
> It wonderful; I call it crass.
> —John Updike, "Cosmic Gall"

THE CULPRIT

Pauli's elusive particle turned out to be just what Enrico Fermi needed, in 1934, to construct a theory of the weak interaction. Fermi synthesized in precise mathematical terms what was then known. For the next twenty years, theorists tried to improve his theory. But they always presupposed parity invariance, and things never quite fit.

Once parity was known to be violated, theorists were free to write down previously forbidden equations, and a basically correct theory of the weak interaction was formulated in 1957 by Richard Feynman and Murray Gell-Mann, and, independently, by Robert Marshak and George Sudarshan.

With further sleuthing, theorists were able to point their fingers at the elusive neutrino as the culprit "responsible" for parity violation. I will now explain how the neutrino was convicted.

Given a spinning particle moving along a straight line, we can ask whether the spin direction (as defined earlier) is in the direction of motion or in the opposite direction. Physicists say that this particle is left-handed or right-handed, respectively. (Theorists proposed originally that this handedness be referred to as "screwiness," but the editors of the leading American physics journal, *The Physical Review,* insisted on the more dignified terms "helicity" and "chirality." As guardians of the language only slightly less august than the forty "immortals" of the Académie Française, they have won and lost their share of battles in their ongoing struggle with the physics community.)

Chirality, or handedness, can be defined as an intrinsic property only for massless particles. Why can't handedness be defined for a massive particle? Suppose we see a massive particle moving in a certain direction, eastward, say. To an observer moving eastward faster than the particle, the particle would appear to be moving westward. Since handedness describes how the spin direction is aligned with the direction of motion, that observer would disagree with us on the handedness of the particle. In contrast, a massless particle, such as the neutrino, always moves at the speed of light—the maximum speed possible, according to Einstein's theory of relativity. Since no observer can move faster than a massless particle, its handedness is an intrinsic property. For instance, the massless particle of light, the photon, can be either left- or right-handed. If Nature respects parity, this would hold for any particle. But experiments showed conclusively that the neutrino has yet one more bizarre property: It always travels left-handed. The neutrino was caught "red-handed"! For almost thirty years now, experimentalists have searched high and low for a right-handed neutrino, but in vain.

Interestingly the German mathematician Hermann Weyl, whom we will meet again, played with the equation we now use to describe the neutrino way back in 1929, but his work was rejected for physics because it obviously violates parity. Weyl's equation was resuscitated in 1956.

I already mentioned that while physicists were shocked by parity violation, they were even more shocked that Nature violates parity selectively. After the conviction of the neutrino, the selectivity becomes, to some extent, understandable, since the neutrino only participates in the weak interaction (and gravity). But Pauli was still upset. In a letter to Madame Wu, he wrote: "Now, after the first shock is over, I begin to collect myself. . . . What shocks me . . . [now is that] God still appears left-right symmetric when He expresses Himself strongly." Twenty years had to pass before physicists were able to have the first deep understanding of the problem bothering Pauli. It turns out that the other three interactions must have a special structure in order for parity violation to be confined to the weak interaction.

INTO THE ANTIMIRROR

The plot now thickens. In the summer of 1956, Lee and Yang received a letter from Reinhard Oehme, a physicist at the University of Chicago, raising the issue of symmetry between particles and antiparticles. Back in 1929, the brilliant English physicist Paul Adrian Maurice Dirac had startled the physics world by predicting that antiparticles exist. By 1956, the existence of antiparticles was well established. The antielectron (called the positron) and the antiproton had both been discovered.

When a particle meets its antiparticle, they annihilate each other, releasing an enormous amount of energy which then materializes into other types of particles. The annihilation of particles with antiparticles is now routinely observed and studied at accelerators around the world. A beam of antiprotons, for instance, can be produced and made to collide with a beam of protons. The fact that the antiproton exists and is annihilated with the proton had long ceased to be of fundamental interest. Rather, physicists are now interested in what novel types of particles may emerge from the annihilation.

A particle and its antiparticle have exactly the same mass, but opposite charges. Thus, the electron has a negative electric charge, the positron a positive one. Given that the elusive neutrino does not carry electric charge, the curious reader may wonder how one distinguishes a neutrino from an antineutrino. Let me explain one possible way. The positively charged pi meson sometimes decays into a positron and a neutrino. Its antiparticle, the negatively charged pi meson, decays into an electron and an elusive particle that we shall call, by definition, the antineutrino.

Dirac's work indicates that the laws of physics do not favor matter over antimatter. To be precise, I define an operation called charge conjugation, whereby one replaces all the particles participating in a given physical process with their respective antiparticles. For example, under charge conjugation, the collision between two protons becomes the collision between two antiprotons. By definition, charge conjugation does not change the movement of the particles or the way they spin. For example, charge conjugation replaces a left-handed particle with a left-handed antiparticle.

Given a physical process, we apply charge conjugation to it and obtain the so-called charge-conjugated process. If the charge-conjugated process occurs with the same probability as the original process, the physical laws governing it are said to be charge conjugation invariant. This is a long-winded but precise way of expressing the notion that Nature, in Her laws, does not prefer matter over antimatter. (See Figure 3.6.)

We can imagine an antiworld made up of antimatter, just as we can imagine the world inside the mirror. Charge conjugation invariance implies that if a physicist from our world ever gets to compare notes with a physicist from the antiworld, the two should agree completely about physical laws. For example, an anticarbon atom built out of antielectrons, antiprotons, and antineutrons will have exactly the same physical properties as a carbon atom. Everyday objects built out of antiatoms will also have the same properties as the corresponding objects built out of atoms. We are unable to build lumps of antimatter merely because no container is available to hold them.

By 1956, charge conjugation invariance had been verified in numerous experiments. But with the fall of parity, Oehme and others naturally wondered if charge conjugation invariance might also fall.

One can put the issue simply by once again considering the neutrino: Charge conjugation invariance implies that the antineutrino would also be purely left-handed. Experimentalists, therefore, went out and "looked" at an antineutrino. They found that it

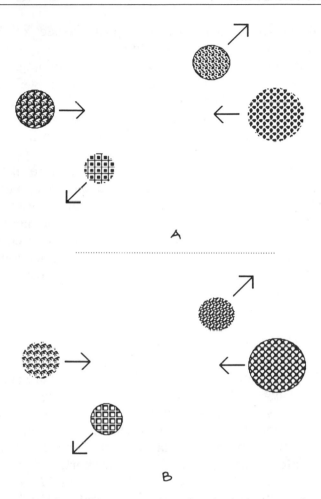

Figure 3.6. (A) A pictorial representation of a physical process in which two particles (the larger circles) collide and are transformed into two other particles (the smaller circles.)

(B) The charge-conjugated process of the process in Figure A: Given a particle, the artist represents the antiparticle by reversing the black-and-white pattern carried by the particle. Charge conjugation invariance states that the processes in Figure A and Figure B occur with the same probability. This states precisely the notion that our world cannot be distinguished from the antiworld.

is in fact right-handed. Weak interaction also violates charge conjugation invariance!

Remarkably enough, the question in this case can also be settled by pure theory. A theorist, in a few lines of mathematical manipulations, could ascertain that charge conjugation invariance is indeed violated by the theory of weak interaction formulated in

1957. This illustrates a most awe-inspiring aspect of theoretical physics. A "good" theory has a life of its own, governed by a secret inner logic. A priori, parity and charge conjugation invariances are logically unrelated issues. Yet, when we incorporate parity violation in a theory (which also has been constructed to respect various established physical facts and principles), the theory comes back to tell us that charge conjugation invariance also fails.

The great theories in physics have far more in them than theorists can imagine at first sight. Indeed, in a philosophical sense, it is misleading to say that a particular theorist invents or creates a certain theory. More properly, he or she merely discovers a theory which, with its myriad mathematical interconnections, has existed for all time. Some of these interconnections may be noticed immediately, but others could lie hidden for decades, or, perhaps—who knows?—forever.

A TANTALIZING PERVERSITY

The simultaneous violation of parity and charge conjugation invariances suggests that if we can construct a magical mirror that not only reflects left and right, but also turns particles into antiparticles at the same time, then the world inside the mirror may be governed by the same physical laws as ours. In other words, while Nature violates charge conjugation, C for short, and parity, P, She may be invariant under the combined operation CP. This possibility is artistically portrayed by the seventeenth-century Dutch painter Pieter de Hooch. The painting of a Dutch courtyard shown in Figure 3.7 is not invariant under reflection alone, but it is approximately invariant if one also turns the woman around, interchanges light and dark, and so forth. The twentieth-century Dutch painter M. C. Escher has fascinated physicists with his geometrical paintings, invariant under reflection followed by interchanging light and dark. (See Figure 3.8.)

Faced with the breakdown of P and C, physicists could at least take a modicum of comfort in the belief that CP is not violated. But even that "security blanket" was to be yanked away several years later. Oehme, collaborating with Lee and Yang, worked out possible experimental tests of CP invariance. In 1964, a team of experimentalists from Princeton University, led by Val

Figure 3.7. Pieter de Hooch, "Courtyard of a House in Delft," 1658 *(Courtesy National Gallery, London):* This painting reminds me of the *CP* (charge conjugation and parity) operation. The woman on the right faces us while the woman on the left shows us her back. The lighted figure of the woman on the right is emerging from a dark background while the dark mass of the woman on the left is going into a lighted background. *(See also* Figure 3.6.)

Fitch and James Cronin, announced that they saw Nature violating *CP*. At that time, I had started my undergraduate studies at Princeton, and I recall that one evening a professor gathered a group of us together and told us the news. Everyone was excited and shocked that Nature had once again been caught committing an impropriety. That Nature could be so tantalizingly perverse per-

Figure 3.8.   M. C. Escher, "Study of Regular Division of the Plane with Birds," 1938. *(Courtesy M. C. Escher Heirs c/o Cordon Act-Baarn, Holland)*

haps contributed to my decision to study physics instead of art history.

The landmark experiment of Cronin, Fitch, et al. involves examining the decays of a certain species of K meson, a strange particle. An analysis based on the principles of quantum mechanics predicts that if CP invariance holds, the K meson should decay into two pi mesons. The K meson does decay into two pi mesons, as predicted by CP invariance—most of the time. Those patient experimentalists from Princeton noticed that once in several thousand decays a K meson would decay into *three* pi mesons!

As a theoretical physicist, I am not particularly interested in the details of how the K meson decays—in itself, that is no more interesting than, say, the behavior of chemicals with unpronounceable names. The deviation of Nature from what is expected of Her interests me.

Parity violation, while shocking enough, is maximal and absolute, in the sense that every neutrino ever "seen" is left-handed, never right-handed. Nature violates parity with a clean-cut finality that some theorists ultimately find comforting. In disturbing contrast, Nature appears to be saying, lazily, that once in a long while She will throw in a bit of CP violation just to confound those nosy physicists.

Since 1956, parity violation has been observed in every single process involving the weak force. Yet after twenty years of trying, experimentalists have not found *CP* violation in any process other than the decay of *K* mesons. Perhaps we will have news soon.

Meanwhile, theorists have failed to agree on a theory of *CP* violation. In contrast, an accepted theory incorporating parity violation was already formulated by 1957, as I mentioned earlier. Many theorists, myself among them, think that *CP* violation is due to a new interaction, weaker than the weak interaction. Others disagree.

While a deep understanding of *CP* violation is lacking, an intriguing consequence has surfaced from cosmological considerations. Several years ago, theorists managed to produce a script in which the universe could have started out empty and evolved to contain matter, and, by extension, us humans. This is a very interesting story in its own right, a story that we will pick up later. Suffice it to say here that for the scenario to work, obviously the laws of physics, at some level, must favor matter over antimatter.

WHAT SHE PLEASES

The reader probably wants to ask why Nature violates parity. Well, who knows? Nature, like the gorilla in the classic joke, does what She pleases.

I am among a group of physicists who still feels, deep down, that Nature really should respect parity. *The New York Times* editorial on parity was entitled "Appearance and Reality." Did the editorial writer mean to imply by the title that, in the esteemed opinion of the newspaper, Nature only appears to violate parity? Perhaps the editorial writer made a Faustian deal and knew more than he or she let on!

The Austrian philosopher and physicist Ernst Mach (1838–1916) once gave a beautiful illustration of appearance and reality. Mach, an extreme positivist who had the distinction of being attacked by Lenin, wrestled with the philosophical problems posed by physics, and his musings deeply influenced Einstein, among others. Mach wrote that, as a child, he was profoundly disturbed to learn that upon the passage of an electric current in a wire placed along a compass needle, the needle would turn. (See Figure 3.9.)

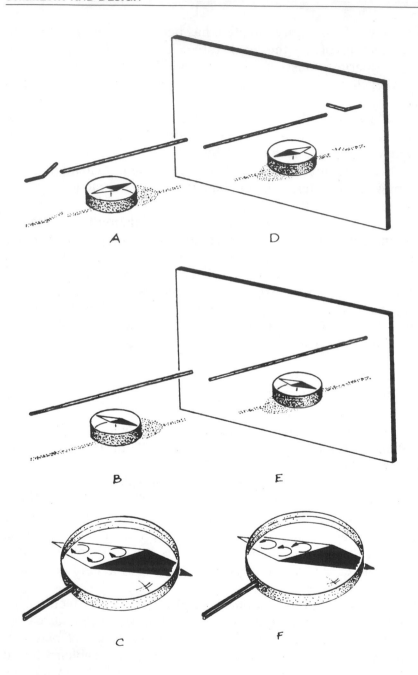

Figure 3.9. The phenomenon which profoundly disturbed the young Mach: In Figure A an electric wire is placed over a compass in the direction of the compass needle. The two ends of the wire are connected to a battery (not shown). The open switch indicates that the wire is not carrying any electric current. In Figure B, the switch is closed and a current flows along the wire, away from the mirror. Mach was outraged to learn that the magnetic field produced by the electric current would cause the compass needle to turn. Believing that Nature could not favor right over left or vice versa, he argued that the needle should refuse to move, since by moving the needle would be manifesting Nature's preference for one side over the other. The conundrum is underlined by considering what would happen in the mirror world. (See D and E.) Many compass needles are traditionally marked with two different colors so that the north-seeking end can be distinguished from the south-seeking end. For the sake of definiteness, the artist painted the south end white. If in our world the current flows away from the mirror, then in the mirror world the current would be flowing into the mirror. Stand facing the mirror so that the current is flowing toward you; you see the south end of the compass needle swing to your left (B). However, your mirror reflection would see, with the current flowing towards him, the south end of the compass swinging to his right (E).

This shocking violation of parity is however only an illusion. If we examine the compass needle in Figure B microscopically, as the artist has indicated with the magnifying glass (C), we would "see" that the magnetism of the compass needle is due to a large number of electrons all spinning in the same direction, clockwise when viewed from above, as the artist had indicated with the three curled arrows. Which end is north and which end is south is determined by the spin direction of the electrons. The paradox is now resolved by examining the compass needle in the mirror world (E). Because of the mirror reflection, the electrons in that compass needle are spinning anticlockwise when viewed from above, as the artist had indicated in Figure F. Thus, in the mirror world, the end painted white would actually be the north end. We were fooled by the black-and-white markings into taking the north end as the south end! In other words, in the last sentence of the preceding paragraph, the word "south" should be replaced by "north." The physicist in the mirror sees the *north* end of the compass needle swinging to his right.

Will a deeper understanding reveal the parity violation presently observed in the weak interaction also to be an illusion?

Since the experimental arrangement is completely symmetrical, the needle should not prefer one side or the other, but should simply decline to budge. The young Mach was upset because parity appeared to be violated. However, if we examine a magnet microscopically, we see that it is simply a piece of metal in which electrons happen to be all spinning in the same direction. The direction of spin points to the "north" end of the compass needle. Suppose now we place a mirror perpendicular to the wire and climb into the mirror world. We would see that the direction of spin inside the magnet is reversed, and thus the compass needle in the mirror has its north and south interchanged. A careful study of Figure 3.9 shows that electromagnetism in fact respects parity. The parity violation that disturbed the young Mach is only an illusion.

Weyl, and later Yang, seized upon Mach's intellectual trauma as an analogy to suggest that with a deeper understanding we may realize that Nature does respect parity. I believe that they are right. Indeed, several theorists have already proposed plausible schemes in which Nature, at a deeper level, will show that She is impartial toward left and right. We will discuss some of these proposals in a later chapter.

When we look at an oriental carpet, any left-right symmetry is immediately apparent. We will now go on looking for subtler symmetries in the tapestry that Nature has woven for us. As with art appreciation, the subtler the symmetry, the greater our pleasure.

II

EINSTEIN'S LEGACY

Marriage of Time and Space

For almost three hundred years, physicists had only reflection and rotation invariances as examples of symmetry. Since these two symmetries could be immediately perceived, physicists generally did not make a big fuss over symmetry as a fundamental concept. Indeed, the notion of symmetry was rarely stated explicitly before the twentieth century.

In 1905, Einstein proposed the special theory of relativity and revolutionized our understanding of space and time. I like to regard Einstein's theory as the very first instance in which physics uncovered a symmetry that Nature has taken some pains to conceal. As we will see in this chapter, it took a considerable amount of connoisseurship to recognize the symmetry of relativity in Nature's design.

The layman interested in physics has long been captivated by the astounding, almost science fiction–like conclusions reached by Einstein. In this book, however, I draw a sharp distinction between the physical consequences and the intellectual foundation of a physical theory.

The intellectual foundation of Einstein's theory consists of a profound appreciation of the power of symmetry, and it was on this foundation that the actual physical consequences of the theory were worked out.

Yes, indeed, the physical consequences of Einstein's ruminations are astounding beyond belief: Mass is equivalent to energy and time is married to space. Who could but be amazed? It is natural that most popular expositions of Einstein's work emphasize these strange features. As a result, such treatments fail to highlight what I consider to be Einstein's truly magnificent intellectual legacy, namely, his views on symmetry. He is the one who groomed symmetry for its starring role in modern physics.

GENTLY DOWN THE STREAM

The concept of relativity did not originate with Einstein. Rather, it is deeply rooted in our everyday perceptions of motion and, as such, had already been built into Newtonian mechanics.

Bishop Berkeley (1685–1753), the fellow who wondered whether a falling tree deep in the forest made noise if no one was around to hear it, worried about how one could say that an object was moving unless another object was present. Anyone who has ridden trains probably has experienced what the bishop had in mind. Sitting in a train in a dark station, on occasion I have been so absorbed in a magazine that I did not notice if the train had started moving. Suddenly, looking out the window, I see the train next to my train slowly gliding out of the station. But is that train moving, or is mine? In the absence of engine noise or any jerking motion, I am eerily confused. I desperately look for some architectural elements, a column perhaps, or a stationmaster standing on the platform. Similar experiences occur in other common situations, such as in a plane taxiing on a runway, or in a boat drifting smoothly downstream. In a four-line, lyrical poem describing a boat trip on a windy day, the Sung dynasty poet Yu-yee Chen (1090–1138) wrote:

> The boat looks red amidst the dancing flowers,
> A hundred li of elms in a half-day's wind,
> Reclining I watch the clouds standing still,
> Not knowing the clouds and I are both traveling east.

In this case, the poet's conception of motion agrees with the physicist's. To the poet, the clouds can quite accurately be described as standing still.

Bishop Berkeley's point is that when we say an object is moving, we really mean that the distance between it and another object is changing with time. The train passenger knows that he is moving when he sees the stationmaster receding.

In an inflationary economy, we are interested in whether our income is increasing relative to that of our neighbors. If every-

Figure 4.1. A contemporary artist's interpretation of a twelfth-century explanation of the relativity of motion.

body's income increases at the same rate, then no one is actually advancing economically. If we are in the happy situation in which our income is rising relative to our neighbors', our neighbors could rightly feel that their incomes are falling relative to ours. Thus, the bishop wondered, if we say an object is moving because its distance from another object is changing, why can't we say that the other object is moving in the opposite direction? After all, the distance between two objects is defined without favoring one or the other of the two objects. The train passenger could say, with perfect justification from a physical-philosophical point of view, that the stationmaster, together with the platform and the entire earth attached to it, is moving in the other direction.

From a practical point of view, it is, of course, more convenient to say that the train is moving. But we must keep in mind that the "common sense" description is more convenient merely because the earth is so much larger than the train. Nowadays, we can watch in our living rooms an astronaut grappling with a disabled satellite. In this situation, the astronaut and the satellite are not too disparate in mass. When the astronaut gives the satellite a push, we see him drifting away from the satellite, but we could say equally well that the satellite is drifting away from him.

In our everyday perception of motion, we are cued by engine noise and jerkiness of motion. But imagine traveling in the distant future in a starship deep into space, far away from any galaxy. Engineering has reached such a state of perfection that there is no engine noise whatsoever. We look out the window and see only the darkness of space. How can we tell if we are cruising steadily or sitting at rest? According to Bishop Berkeley, we cannot; absolute motion cannot be defined.

Now suppose we look out of the window and see another starship approaching. Are we moving toward this starship, or is it moving toward us? It is not possible to tell, so the question is meaningless. We can only say that our starship is moving relative to the other one.

In a sense, we are traveling on a starship right now. Our entire galaxy is known to be moving toward the Virgo cluster of galaxies at a couple of hundred kilometers a second—faster, literally, than a speeding bullet—yet we hardly feel it. There is no engine noise to speak of. But are we moving or is Virgo moving toward our Milky Way?

That the relative motion is at constant velocity is essential. As soon as the starship pilot "steps on the gas," we will know that we are speeding up. We experience this point almost daily. When a car suddenly speeds up, passengers are thrown back.

All of this was well understood by Galileo. Instead of starships, he spoke of good old-fashioned sailing ships.

RELATIVITY OF MOTION AS A SYMMETRY

The impossibility of defining absolute motion can be seen as the manifestation of a symmetry known as relativistic invariance. In the same way that parity invariance tells us that we cannot distinguish the mirror-image world from our world, relativistic invariance tells us that it is impossible to decide whether we are at rest or moving steadily. To avoid possible confusion later, let us pin down a precise definition of this concept.

Consider two observers moving relative to each other, per-

Figure 4.2. On a train moving smoothly at 30 feet per second, a stoker throws a piece of coal forward at 10 feet per second. Standing on the ground, we see the coal flying forward at 40 feet per second.

fectly smoothly, at a velocity that does not change with time. We will refer to this type of movement as motion at constant velocity. As an example, we may picture a train moving smoothly at thirty feet per second relative to the platform. Suppose a passenger sitting at the back of the carriage tosses a ball to the front of the carriage at ten feet per second. How fast does the ball travel, as measured by the stationmaster on the ground? Most of us intuitively would find it obvious that to the stationmaster the ball moved forward at $30 + 10 = 40$ feet per second. Generally, for every physical quantity measured by two observers in relative motion at constant velocity, be it the velocity of a ball or the temperature of a cup of coffee, a formula relates the two measurements. In our example, if the velocity of the ball, as measured by the passenger and the stationmaster, is denoted by v and v', respectively, and if the velocity of the train relative to the station is denoted by $u,$ then we conclude that $v' = v + u.$ (For the specific numbers given previously, u equals thirty feet per second, v, ten, and v', forty.) The collection of all such formulas, relating velocity, energy, momentum, temperature, and so forth, as measured by two different observers, is known as a Galilean transformation.

Now, suppose the two observers are physicists who want to determine the laws of physics. For instance, the passenger and the stationmaster could both decide to determine the law governing the motion of the ball. Relativistic invariance says that two observers in relative motion at constant velocity must arrive at the same physical laws, in spite of the fact that they differ in their measurements of various physical quantities. Thus, in our example, while the passenger and the stationmaster differ in their perception of how fast the ball is moving, they must both arrive at Newton's law.

This definition of relativistic invariance expresses in precise terms that it is physically impossible to tell which one of two observers in relative motion is actually moving. If the physical laws observed by the two observers in relative motion at constant velocity, were not the same, then Nature would be distinguishing between the two observers.

In the preceding chapters, I spoke of observers with their heads tilted relative to each other, and of an observer "outside" the mirror, another "inside" the mirror. In this chapter, I speak of observers in relative motion. The basic notion of symmetry is the same in all these cases. The issue of symmetry is whether different observers perceive the same structure of physical reality.

ON THE ELECTRODYNAMICS OF MOVING BODIES

By emphasizing the bizarre aspects of Einstein's theory, some popular expositions end up making relativity sound more mysterious than it really is. In fact, relativity represents a logical, almost inevitable progression of ideas flowing out of the nineteenth-century understanding of electricity and magnetism.

A proper understanding of Einstein's theory is impossible without some understanding of its roots in electromagnetism. After all, Einstein, with a modesty unheard of nowadays, titled his epoch-making paper simply "On the Electrodynamics of Moving Bodies." Thus, I propose to show the reader a flashback of the development of electromagnetic theory.

FROGS AND LODESTONES

Electric and magnetic phenomena had been known for a long time. The magic of amber and lodestone fascinated the ancients. After a piece of amber is rubbed on a furry substance, it can pick up bits of hair and paper. (Children know that a plastic comb works just as well.) As for the mysterious lodestone, we now know it to be a naturally magnetized piece of iron ore. The ancient Chinese discovered enough of its properties to be able to construct the magnetic compass.

William Gilbert (1544–1603), the royal physician to Elizabeth I, was the first to recognize the distinction between electric and magnetic forces. His work cleared up a great deal of confusion. After Gilbert, electric and magnetic phenomena were studied separately.

Progress was slow and sporadic. For instance, Charles François de Cisternay du Fay, a swashbuckling courtier of Louis XIV and a leading scientist of that era, would amuse the court by electrifying people to elicit sparks from their fingers. (Physicists had more fun in those days!)

In 1785, Charles Augustin Coulomb (1736–1806) determined that the electric force between two electrified objects varies inversely as the square of the distance separating them. This quantitative description of the electric force became known as Coulomb's law.

Further progress had to wait for the "accidental" discovery made by anatomist Luigi Galvani (1737–1798) while dissecting a frog in that momentous year, 1789. He found that when two different metals came into contact with the frog's legs, the frog twitched. We know now that an electrical impulse went through the frog's legs: Animals can produce electrical currents. Indeed, our nerves and muscles are controlled by electrical impulses.

The biologist Count Alessandro Volta (1745–1827) then took the crucial step of separating biology from physics. Volta demonstrated that the electricity produced does not depend on the frog, which can be ignominiously replaced by a chemical fluid. An arrangement of metal plates immersed in a suitable chemical bath will produce electricity. And so the battery was born.

With batteries providing a controllable flow of electricity, physicists could experiment with electricity and magnetism systematically. In 1819, Hans Christian Oersted (1777–1851) discovered, supposedly by accident, that when an electric current flows through a wire, a compass needle in the vicinity moves. (See Figure 3.9 on page 46.) As we have seen, the apparent violation of parity in Oersted's phenomenon deeply disturbed the young Mach. An electric current can generate a magnetic field! Electric and magnetic phenomena are related, and physics was soon to have a new term: electromagnetism.

Oersted's astounding discovery opened a most exciting era in which science and technology evolved together at a dizzying pace. Imagine, one could telegraph across the Atlantic less than a hundred years after Galvani's frog gave up his life. Within a fifty-year period, roughly between 1825 and 1875, such inventions as telegraphy, the electric motor, and the electric generator, all of which provide the foundation for the modern world, were developed.

Let us return to physics. Two mysterious phenomena were found to be related, and many new questions were raised.

In Oersted's experiment, an electric current moves a magnet. What about the reverse? Holding the magnet fixed, could one cause the wire carrying the electric current to move? The answer is yes. Exploiting this phenomenon, one can construct the electric motor.

If electricity can generate a magnetic force, as Oersted had shown, can magnetism produce electricity? If one moves a magnet around a wire, would that cause an electric current to flow? The answer, once again, is yes, a moving magnet generates electricity.

And so the race was on. We can easily picture Victorian physicists in their laboratories, with wires, magnets, and voltaic cells (primitive batteries), in feverish excitement trying all possible configurations as Nature revealed Her secrets one after another.

MAY THE FIELD OF FORCE BE WITH YOU

Of the many eminent experimenters of the era, Michael Faraday (1791–1867) is often regarded as the greatest. While Faraday's genius manifested itself in the laboratory, he also introduced into theoretical physics the important and fruitful concept of a "field of force," or "field" for short.

Unlike most physicists until his time, Faraday did not come from a comfortable background. Born into almost Dickensian poverty, Faraday started as a bookseller's errand boy, later promoted to an apprentice. While rebinding a set of the *Encyclopaedia Britannica,* he became spellbound by an article on electricity he chanced upon. In Victorian London, educational lectures were often given to the public, typically for a charge of one shilling a lecture, a fee the young man was hard put to come up with. Fortunately, the famed Sir Humphrey Davy started to give free lectures at the newly founded Royal Institution. They were highly popular. The educated public was intimately interested in science,

Figure 4.3. Michael Faraday (drawing after an original portrait). The field of force is represented by arrows indicating the direction a charged particle would move if placed at the location of the arrow.

and electricity was, well, electrifying the public. (This fine tradition of free lectures has persisted to this day in many countries, and most physics centers I know of can boast of one or two strange wild-eyed characters in regular attendance at seminars and colloquia.)

Faraday, who attended the lectures religiously, eventually approached Davy. As luck would have it, Davy was at that very moment in need of a laboratory assistant. Furthermore, he was to go on a tour of European science centers a few months later and offered to take Faraday along. So Faraday did end up with an education to be envied. The Dickensian scenario was complete, however; Lady Davy was a horrid snob who insisted that Faraday eat with the servants and generally made life unpleasant. He was often reduced to performing the tasks of a valet. But it was an exciting trip, scientific and otherwise; the Napoleonic wars were in full swing, and, as "enemy scientists," they had to travel on "safe-conduct" through the lines.

Davy's young assistant quickly established himself, making discoveries one after another and outshining his mentor. Jealousy is a powerful human emotion and unpleasantness soon developed between the two men. Among other things, Sir Davy tried to block Faraday's membership in the Royal Society, but in vain. At the height of his career, Faraday was showered with honors. The humble apprentice was to refuse a knighthood as well as the presidency of both the Royal Institution and the Royal Society. Even Davy admitted that of all his discoveries, Faraday was the best.

But what is this field of force discovered by Faraday and now known to every child who has seen films on interstellar warfare?

In our everyday experiences, we tend to think of a force being exerted only when contact is made between material bodies, as when we push open a door. Newton's law of gravitation had already introduced the notion that a force could act at a distance. But this idea of "action at a distance" deeply troubled many thinkers. At any moment in time, the earth has to "know" instantaneously the sun's position and to "feel" the appropriate force. The phenomenon of electromagnetism demonstrated this apparent action at a distance even more dramatically. That magnets would act on each other while separated by empty space is most alluring to children, and to physicists as well.

Like many of his predecessors and contemporaries, Faraday grappled with this philosophical problem and finally reached the following picture.

He proposed that an electric charge produces around it an electric field of force. When another charge is introduced into this electric field, the field acts on this charge, exerting on it a force in accordance with Coulomb's law.

The important point is that this electric field is to be thought of as a separate entity: The electric field produced by an electric charge exists, regardless of whether another charge is introduced to feel the effect of the field. Similarly, one envisages a magnetic field produced by a magnet or an electric current. Thus, Faraday introduced an intermediary: Two charges do not act "directly" on each other but they each produce an electric field that, in turn, acts on the other charge.

A pragmatic physicist was apt to dismiss all this as just talk that did not advance our knowledge one whit. Faraday's notion does not explain Coulomb's law in any sense; rather, it appears to be merely another way of describing Coulomb's law. Faraday supposed the strength of the electric field produced by a charge decreases as one moves farther away from the charge, in such a way as to reproduce Coulomb's law.

But this view missed the point. The real content of Faraday's picture, as it turns out, lies in the fact that the electromagnetic field not only can be thought of as a separate entity, it *is* a separate physical entity. Physicists were to learn, for example, that it makes perfect physical sense to talk of the energy density in an electromagnetic field. As we will see, the notion of a field would bear fruit in the hands of Scotsman James Clerk Maxwell (1831–1879).

TELECOMMUNICATION, FRENCH PHILOSOPHERS, AND PIGEONS

Because of his up-from-rags background, Faraday had a self-admitted blind spot—mathematics—and he was unable to transcribe his intuitive notions into precise mathematical descriptions. Just the opposite, Maxwell, scion of a distinguished family, received the best education that his era could provide, and was thereby able to achieve the grand mathematical synthesis of electromagnetism. But before he was to begin his investigations, Maxwell made a resolution: "To read no mathematics on the subject [of electricity] till I had first read through Faraday's *Experimental Researches on Electricity.*"

Some young contemporary theoretical physicists are so enamored of mathematics that they might well take heed of Maxwell's comment. Indeed, Maxwell was to consider Faraday's deficiency an advantage. He wrote:

> Thus Faraday, with his penetrating intellect, his devotion to science, and his opportunities for experiments, was debarred from following the course of thought which had led to the achievements of the French philosophers, and was obliged to explain the phenomena to himself by means of a symbolism which he could understand, instead of adopting what had hitherto been the only tongue of the learned.

By "symbolism," Maxwell was referring to the notion of field, actually called "lines of force" by Faraday. Earlier, Maxwell had said, "the treatises of [the French philosophers] Poisson and Ampère [on electricity] are of so technical a form, that to derive any assistance from them the student must have been thoroughly trained in mathematics, and it is very doubtful if such a training can be begun with advantage in mature years." Indeed, the pace at which sophisticated mathematics has been introduced into theoretical physics in recent years is such that many physicists "in mature years" would share heartily the sentiments expressed by Maxwell.

The American school of theoretical physics by tradition has stressed physical intuition, at the expense of what is sometimes referred to as "fancy shmancy mathematics." I will refrain from exploring the historical and sociological origins of this emphasis, which has been, at once, this philosophy's strength and its weakness. Generally speaking, European physicists receive a much more vigorous training in contemporary mathematics than their American counterparts. The French philosophers, now referred to as the French physicists, still are regarded by many Americans as incomprehensibly mathematical. Of course, what is considered fancy by one generation is often thought basic by the next. The mathematics used by Poisson et al. now looks like child's play and is familiar to any undergraduate student of physics.

LET THERE BE LIGHT, BUT WHAT IS LIGHT?

At midcentury, Maxwell took stock of the accumulated knowledge about electromagnetism. The end result of a century of arduous experimentation had been distilled and summarized already in various laws, named after various investigators. Maxwell put it all together in four mathematical statements, known ever since as Maxwell's equations. The equations specify how the electromagnetic field varies, in space and in time. For instance, one equation states how the electric field varies in space, in the presence of a magnetic field that is varying in time. It expresses, in concise mathematical terms, Faraday's law of induction: By moving a magnet around a wire, one produces an electric field that pushes charges forward in the wire, generating a current. Another equation specifies how the electric field around a charge decreases with distance away from the charge, thus reexpressing Coulomb's law.

Consider a detective faced with a complicated criminal case. He has spent weeks gathering testimonies. Finally, he sits down to check whether the testimonies are mutually consistent. Hmm, the butler couldn't be telling the whole truth. But . . . aha! If the butler actually said 12 A.M. instead of 12 P.M. then everything falls into place. So, too, Maxwell sat down and asked if the four equations he had written were mutually consistent. Hmm, this one can't be right—it contradicts the other three! Remarkably, Maxwell then noticed that by modifying the offending equation slightly, he could bring the four into harmony.

Armed finally with the correct equations, Maxwell was able to go further. In a flash of insight, he made one of those truly amazing discoveries in physics: the existence of electromagnetic waves. Roughly speaking, if we have in a region of space an electric field changing in time, then a magnetic field is produced in the neighboring space. Its very production means that this magnetic field is also changing in time—and *it* generates an electric field. Thus, like a ripple on a pond spreading from a dropped pebble, an electromagnetic field propagates out in a wave, undulating between electric and magnetic energy.

Maxwell was able to calculate precisely from his equations the speed of this electromagnetic wave. By his time, the speed of light had been measured quite accurately, both by terrestrial experiments and by astronomical observations. The value obtained theoretically for the speed of his electromagnetic wave coincides closely with the measured speed of light! And thus Maxwell proclaimed that the mysterious phenomenon of light is just a form of electromagnetic wave. In one stroke, optics as a field of physics was subsumed under the study of electromagnetism.

The laws of optics, wrested from Nature by physicists starting with Newton and Huygens, would be derived entirely from Maxwell's equations. Before, human vision had been limited to a narrow window in the electromagnetic spectrum; after, all forms of electromagnetic waves were ours to exploit. Telecommunication was born.

Physicists have often used the birth of telecommunication to illustrate the importance of funding basic research. They easily can imagine the Royal Navy official charged with allocating funds to improve communication deciding it would be folly to support these strange types fooling around with wires and frogs' legs in their gloomy laboratories. Obviously, he might have reasoned, the money would be better spent on breeding a speedier strain of carrier pigeon.

Maxwell's discovery demonstrated conclusively the physical reality of the field and its claim to a separate existence. Indeed, the space around us is literally humming with packets of electromagnetic field hurrying hither and yon. The notion of field has grown from a glint in Faraday's eyes to be all-encompassing. In recent decades, physicists have come to the view that all physical reality is to be described in terms of fields, an idea we will come back to later. It is interesting how this concept originated in the vague philosophical unease physicists felt with the action-at-a-distance hypothesis.

THE BIG QUESTION

Let us return now to Einstein and relativistic invariance. It is the end of the nineteenth century, and physicists are justifiably proud of their success in understanding electromagnetism. The stage is set for Einstein and others to ask the $64,000 question. We have seen that Newton's theory of mechanics is invariant under the Galilean transformation, Galilean invariant for short. Is Maxwell's theory of electromagnetism also Galilean invariant?

To answer this question, let us go back to the train moving

smoothly at thirty feet per second. Suppose that the passenger, instead of tossing a ball forward, shoots a beam of light forward. Denote the speed of light, as measured by a physicist on the train, by the letter c. Galilean transformation tells us that the speed of light, measured by the physicist on the ground, ought to be $c + 30$ feet per second.

But wait! Recall that Maxwell was able to calculate the speed of light using his equations.

These equations incorporated the measurements of Oersted et al. For instance, one of these might measure the strength of the magnetic field generated by an electric field, varying at such and such a rate. But a physicist performing Oersted's experiment on the train should arrive at precisely the same result as a physicist performing it on the ground, since otherwise, the two physicists would perceive two different structures of physical reality. These two experimentalists can now appeal to their respective theoretical colleagues to perform Maxwell's calculation of the speed of light. If the two theorists are both competent, they should arrive at the same answer. Thus, if Maxwell's equations are correct, the speed of light, as measured by the observer on the train and on the ground, should be exactly the same! This strange behavior of light indicates that physics cannot be Galilean invariant.

Maxwell's reasoning forces us to a conclusion in violent disaccord with our everyday intuition: The observed speed of light is independent of how fast the observer is moving. Suppose we see a photon whizzing by and decide to give chase. We get into our starship and gun the engine till our speedometer registers nine-tenths the speed of light. But when we look out the window, to our astonishment we still see the photon whizzing by at the speed of light. The photon would make an unbeatable track star.

The key point is that the speed of light is an intrinsic property of Nature, deriving from the way an electric field varying in time generates a magnetic field and vice versa. By contrast, the speed of the tossed ball in our example depended on the muscular prowess and inclination of the tosser.

EINSTEIN AND TIME

Physics is not Galilean invariant. Now what?

To proceed, recall that symmetry is composed of two logically distinct components; invariance and transformation. To say that physical laws are invariant, we must specify the transformation that leaves the physical laws invariant. For rotation symmetry, the transformation involved is a rotation. For reflection symmetry, the transformation involved is a reflection. In discussing reflection and rotation symmetries, there is no question as to what the corresponding transformations are, and so we tend not to emphasize these two distinct components. Strictly speaking, we should say, for example, that physics before 1956 was believed to be invariant under that transformation commonly known as reflection.

In our discussion of relativistic invariance we have assumed that the relevant transformation is the Galilean transformation. When faced with the conclusion that electromagnetism is not relativistic invariant under the Galilean transformation, a lesser physicist might have been tempted to abandon the notion of relativistic invariance. But this position appeared untenable, since the speed of light is invariant. Faced with this confusing and paradoxical situation, Einstein boldly insisted that physics must be relativistic invariant, a position that forced him to abandon Galilean transformation. The specific transformation is to be jettisoned, not the notion of relativistic invariance.

The boldness of Einstein's position becomes apparent if we reflect on the fact that the Galilean transformation of velocity is based on our fundamental understanding of the nature of time. Let us return once again to the train. To say that the train is traveling at thirty feet per second, we mean that when one second has elapsed for the stationmaster the train has moved forward by thirty feet. To say that the ball is tossed forward at ten feet per second, we mean that when one second has elapsed for the passenger, the ball has moved forward by ten feet, relative to the tosser sitting in the train. Newton, and everybody else, made the unspoken but eminently reasonable assumption that when one second has elapsed for the passenger, precisely one second has also elapsed for the stationmaster. Time thus conceived is referred to as absolute Newtonian time. Given absolute Newtonian time, the stationmaster would then conclude that during the passage of one second, the tossed ball has hurtled forward through space by $30 + 10 = 40$ feet.

Thus, Einstein was forced to throw out the cherished notion

of absolute time. Different observers in relative motion at constant velocity perceive the passage of time differently.

Since train speeds are small compared to the speed of light, passengers hardly notice the failure of absolute time. In particle accelerators, however, where subnuclear particles move at speeds close to that of light, Einstein's revolutionary conception of time is now verified every day. By working out the mathematical description of relativistic invariance, Einstein was able to predict that subnuclear particles moving at high speeds would be measured by the experimenter as living longer than those sitting still in the laboratory. When we say that a particle is moving at high speed, we could just as well say that the experimenter is moving at high speed relative to the particle. How long the particle lives before it disintegrates is an intrinsic property of that particular species of particle, but how long the particle lives according to the *experimenter's* clock depends on how fast the experimenter is moving relative to the particle: The faster the relative motion, the longer the measured lifetime.

Unfortunately, or perhaps fortunately, Einstein's theory does not offer a path to longevity. The lifetime of the train passenger is measured as longer by the station clock, but the lifetime experienced by the passenger, that is, as measured by the clock in the train, remains the same. In fact, since the very notion of relativity insists that neither the passenger nor the stationmaster has a status more special than the other, the lifetime of the stationmaster is also observed by the passenger as longer. Each perceives the other as having lived longer!

In view of this strange property of time, it is convenient and natural to introduce the notion of "proper time." Imagine that every particle in the universe carries with it its own clock. Proper time of a given object is defined as the time recorded by the clock carried by that object. Clearly, proper time is the only intrinsically significant measure of the passage of time. For example, when physicists list the lifetime of a certain subnuclear particle in a textbook, they are referring to the particle's lifetime as measured by its own clock, not the experimenter's. To refer to the particle's lifetime as measured by the experimenter, one would have to specify the relative velocity between the particle and the experimenter, a physical quantity not intrinsic to the particle and hence variable from experiment to experiment.

As I mentioned before, for a given interval, proper time is always less than the time measured by another observer. Physicists say that time is dilated by movement. For each of us, our own perception of time is always less than that of anyone else. In our example, the train passenger perceives his own lifetime to be shorter than what the stationmaster perceives it to be. The higher the relative velocity between the observer and the observed, the larger is the ratio of observed time to proper time. For a photon, cruising at the ultimate speed limit, the passage of eternity is but an instant. Such is the hot-rodder's dream! In fact, the clock carried by a photon is stuck; a photon's proper time never changes.

The bizarre behavior of time astounded and fascinated the public. Perhaps the best-known poem about the relativity of time is a limerick penned by A. H. R. Buller, published in the comic magazine *Punch:*

> There was a young lady named Bright
> Whose speed was far faster than light;
> She went out one day,
> In a relative way,
> And returned the previous night.

The poet indulged in a certain amount of license: Einstein's equations specifically forbid returning before one has left! The best one can do, if one is a photon, is to return at the same instant of *proper* time as one departed.

TIME AND MOTION IN THE SIXTEENTH CENTURY

Miss Bright's strange trip reminds me of another famous trip, that of Magellan, in which time also took an unexpected turn. After spending three years circumnavigating the globe, Magellan's expedition finally sighted a Portuguese island on Wednesday, July 9, 1513, according to the ship's log. But the landing party was bewildered and perturbed when the islanders insisted that it was actually Thursday. This phenomenon, now all too familiar to travelers with jet lag, puzzled the intelligentsia of the time greatly. It is of course merely a consequence of a certain human convention of recording time and has nothing to do with relativity. The proper time experienced by Magellan and by the Portuguese islanders differs only imperceptibly on the human time scale. (Actually, Magellan had been speared in the Philippines, but we follow the academic tradition of giving him credit even though he died in the middle of the project he initiated.)

A NEW TRANSFORMATION OF SPACE AND TIME

Physicists are fond of saying that Einstein merged space and time into "spacetime." To understand what this means, we have to learn how physicists described the strange behavior of time.

In the same way that historians record events, physicists record an event in the physical world by specifying its location in time and space, namely by assigning four numbers, t, x, y, z, corresponding to the event. The time, t, is measured from some mutually agreed-upon event, in much the same way that Western historians commonly used the birth of Jesus Christ as the reference point in time. The other three numbers, x, y, z, specify the location of the event in three dimensional space, as measured from some agreed-upon reference point.

In our example, a given event would be recorded by the passenger as occurring at t, x, y, z, and by the stationmaster at t', x', y', z'. To specify the transformation laws of space and time is to supply the mathematical formulas relating (t, x, y, z) and (t', x', y', z'). Thus, the Galilean transformation asserts that $t = t'$, that time is absolute. As we have seen, Einstein was forced to throw out Galilean transformation. But then his insistence on relativistic invariance makes sense only if he can find another transformation under which physics is relativistic invariant.

At this point, it is a straightforward mathematical exercise to find the transformation. One simply designs to fit: One demands that the relations between (t, x, y, z) and (t', x', y', z') are such that the speed of light comes out to be the same as measured by the two observers in relative motion at constant velocity. Remarkably, this exercise requires only high school algebra. The determination of how space and time transform turns out to be one of the simplest calculations in the history of physics! Physicists call this transformation the Lorentz transformation, in honor of the Dutch physicist Hendrick Antoon Lorentz (1853–1928).

In the Lorentz transformation, t' is no longer simply equal to t, as is the case in the Galilean transformation, but is given by a mathematical expression involving t, x, y, and z (and the relative

Figure 4.4. How did Einstein derive his famous formula? *(Courtesy Sidney Harris)*

velocity u between the observers, of course). When u is small compared with the velocity of light, c, we expect t' to be approximately equal to t. But in general, the transformed time, t', will depend on t and on the coordinates in space, x, y, and z. The transformed time depends on space. In the same sense, the transformed space depends on time. Thus is time married to space, and space to time. Forever after, physicists were to speak of space and time as one, as spacetime.

REVISIONIST MECHANICS

The discovery of the Lorentz transformation was motivated by one aspect of electromagnetic theory: the peculiar way in which light propagates.

Perhaps not surprisingly, Maxwell's entire theory of electromagnetism turns out to be relativistic invariant under Lorentz transformation, or Lorentz invariant for short. Now comes the all important point: Since mechanics describes the motion of particles in spacetime, our new conception of spacetime obviously dictates that mechanics must be revised so that it, too, is Lorentz invariant. The new subject of electromagnetism forced physicists to modify an older area of physics previously considered absolutely secure. It is not unlike a detective mystery in which a fresh clue, at first apparently unrelated, eventually forces a revision of a hypothesis previously considered established.

Einstein thus proceeded to tinker with Newtonian mechanics and reached an astonishing conclusion about the nature of energy. Everyone knows $E = mc^2$, but how did Einstein know? Basically, Einstein found that in order to make the laws of mechanics invariant under Lorentz transformations, he was forced to modify the definitions of energy and momentum and the relationship between the two.

In Newtonian mechanics, energy is proportional to momentum squared: The faster an object moves, the more energy it has. When the object is sitting still, in other words, when its momentum is zero it has zero energy. Einstein changed this relation so that even when an object is sitting still, it has an amount of energy equal to its mass times the speed of light squared: $E = mc^2$. Since c is so much larger than any commonly attainable velocity, this so-called rest energy is fantastically larger than the Newtonian energy.

The details of Einstein's reasoning are not essential. The important point to appreciate is that by a logical, step-by-step process, the human intellect is able to uncover one of Nature's deepest secrets.

Note that Einstein's formula does not say anything about how to unlock the energy hidden inside mass. That this is in fact possible was demonstrated dramatically in 1938, when two German scientists, Otto Hahn and Fritz Strassmann, managed to split an atomic nucleus. Mankind has learned to release the enormous rest energy contained in matter. The prospect for us is at once liberating and terrifying. Will we use Einstein's formula to reach for the stars, or to annihilate the planet? Will global political leaders have the courage to eliminate or reduce our nuclear arsenal? Will we exploit the rest energy to free the human race from the unspeakable physical labor that many members of the poorer countries continue to endure?

It is misleading to equate Einstein's formula with nuclear energy, as the popular press often does. After all, Einstein arrived at his formula by studying how objects move. Nuclear physics never entered into his reasoning. Einstein's considerations are predicated on the properties of spacetime and so must be universally applicable to all processes. In fact, when we burn a log, if we carefully measure the mass of the log, of the ashes, the cinders, and the glowing hot gas, we will find that a minute amount of mass is missing: It has been converted to energy. Einstein's formula is not only relevant for nuclear energy, it applies to our daily lives as well. We can say that mankind knew all along how to release the energy hidden in mass, although, before 1938, only in minute amounts.

Einstein's discovery is essential for exploring the subnuclear world. We already have seen in the preceding chapter that Pauli was able to deduce the mass of the neutrino by using Einstein's formula. The interconversion of mass and energy is commonplace when particles collide. For instance, the collision of two very energetic protons may produce seventeen extra particles, called mesons, in addition to the two protons. Mass is not conserved; some of the energy of the colliding protons has been converted into the mass of the mesons. Newtonian mechanics is totally incapable of accounting for this sort of phenomenon in which energy is converted into matter. In the everyday Newtonian world, when two billiard balls collide, one of them may shatter into fragments, but we would be mightily surprised to see the scattered billiard-ball fragments accompanied by seventeen pieces of chalk.

The possibility of converting mass into energy also cleared up a long-standing mystery. In the nineteenth century, physicists could not understand how stars could contain enough fuel to keep them burning for eons. We now understand that starfires are fueled by the stars' enormous masses.

INNER CONNECTIONS: THE POWER OF SYMMETRY

Important though Einstein's formula is, it is less interesting from an intellectual standpoint than the power of symmetry. To me, Einstein's formula is part of the libretto of relativity, while the underlying symmetry, the notion of relativistic invariance, provides the music.

The revision of Newtonian mechanics was not up to Einstein; it is dictated by Lorentz invariance. In an earlier chapter, I spoke of the inner life of physical theories, with their labyrinths of secret inner connections to be discovered. The present story illustrates this picture well. When I started to learn physics, I was most impressed by how diverse phenomena, apparently totally unrelated, turn out to be connected at a deeper level. Other sciences are closer to our direct perceptions of the world and thus are perhaps more appealing and more easily appreciated. Hiking through the mountainous regions of the world, I am fascinated by the formations I see, and an understanding of the geological forces at work affords me pleasure. Yet, learning that ancient rivers had gouged out the gorges that I find so magnificent and, well, gorgeous, though it adds to my understanding, does not particularly surprise me. Physics does! That the longevity of stars, the magic of light, the compass needle seeking north, and the frog's leg twitching are all interrelated and controlled by one symmetry principle—now that is a real surprise!

Dirac's 1929 prediction of antimatter provides another stunning example of how symmetry guides physicists to Nature's inner secrets. By the late 1920s, physicists had already discovered the so-called Schrödinger's equation governing the behavior of electrons in atoms. Schrödinger's equation was not Lorentz invariant. However, since atomic electrons move at speeds much less than the speed of light, it was perfectly adequate in describing the known properties of atoms. But Dirac, like Einstein before him, insisted that all of physics must be relativistic invariant. He thus proceeded to make Schrödinger's equation Lorentz invariant. To his surprise, the modified equation, now known as Dirac's equation, possessed twice as many solutions as Schrödinger's. After much puzzlement, Dirac realized that the additional solutions described a particle with properties opposite to those of the electron. The antielectron, now known as the positron, was discovered by Carl Anderson three years later.

The discovery of a symmetry is much more than the discovery of a specific phenomenon. A symmetry of spacetime, such as rotational invariance or Lorentz invariance, controls all of physics. We have seen that Lorentz invariance, born of electromagnetism, proceeds to revolutionize mechanics. And once the laws of motion of particles are revised, our conception of gravity has to be changed as well, since gravity moves particles. In the next chapter, we will see how Einstein tried to make gravity Lorentz invariant and reached conclusions even more astonishing.

DRIVE TOWARD UNITY

Physicists dream of a unified description of Nature. Symmetry, in its power to tie together apparently unrelated aspects of physics, is linked closely to the notion of unity. The story of electromagnetism illustrates well what I mean by the drive toward unity: Electricity and magnetism were revealed to be different aspects of electromagnetism, and optics then became part of electromagnetism.

In high school, I read an old physics book that said physics consists of six parts: mechanics, heat, light, sound, electricity and magnetism, and gravity. In fact, toward the end of the nineteenth century, there were only two fields left in physics: electromagnetism and gravity. The status of the drive toward unity at that time

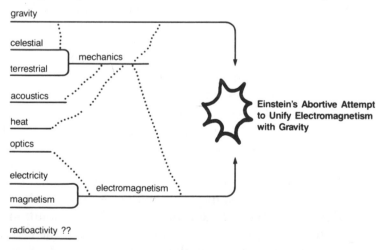

DRIVE TOWARDS UNITY
TOWARDS THE END OF THE NINETEENTH CENTURY

Figure 4.5. Toward the end of the nineteenth century, all of physics was unified into two interactions, electromagnetism and gravity. Radioactivity, which had just been discovered, did not appear to fit. Naturally enough, Einstein tried to unify electromagnetism and gravity. His efforts were doomed because the picture was incomplete: The strong and weak interactions were not yet known.

is shown in Figure 4.5. The drive toward unity may be said to have started with Newton, who insisted that the same laws govern heavenly bodies and celestial objects. Terrestrial and celestial mechanics were unified. Later, sound was recognized as being due to the wave motion of air and it was realized that it could be studied with the concepts of Newtonian mechanics. In the nineteenth century, the mystery of heat was finally understood as due to the agitated motion of molecules. The mechanical interaction between objects, such as that due to friction, was traced to the electromagnetic interaction between the atoms and molecules comprising the objects. If we mean by mechanics the description of the motion of particles, then we may say that mechanics has been subsumed into the other interactions.

I mention in Chapter 2 that as physicists explore Nature at ever-deeper levels, Nature appears to get ever simpler. The story of relativistic invariance exemplifies this remarkable phenomenon. I may surprise the reader by saying that Einsteinian mechanics, once mastered, intrinsically is simpler than Newtonian mechanics. After working with Lorentz invariant equations, I find equations in Newtonian mechanics awkward and malformed. Space and time are not treated on the same footing, and neither are energy and momentum. The equation does not please my eyes, understandably so, since the Newtonian equation is only approximate to the Einsteinian equation. Why should Nature care whether the results of an imposed approximation look pretty? Similarly, recognizing the relativistic invariance of electromagnetism, fundamental physicists now write Maxwell's equations more compactly as one equation. When I was a student, I had to memorize Maxwell equations before every examination. Mmm, let's see, a magnetic field changing in time produces an electric field changing in space—or, is it changing in time? With relativistic invariance, a single equation describes an electromagnetic field changing in spacetime. I find this completely symmetrical equation as easy to remember as the shape of the circle.

Intrinsically, advanced physics is simpler than elementary physics—a little secret not often revealed to the layman. Many people are stumped by high-school or college physics because they are presented with misshapen phenomenological equations having little to do with Nature's intrinsic essence, with Her beauty, Her symmetry, or Her fundamental simplicity.

5

A Happy Thought

INVARIANT STRUCTURE OF REALITY

In 1905, Einstein stunned the physics world with a Lorentz invariant system of mechanics, but his task was not yet done. There remained one well-established area of physics, gravitation, which must be made Lorentz invariant.

Einstein's work on electromagnetism and relativistic mechanics, known as the special theory of relativity, progressed like a hot knife through butter—or at least it seems that way in retrospect. In contrast, the problem of making Newton's theory of gravity Lorentz invariant stumped Einstein. It was only after ten years of incessant struggle that Einstein was finally able to come forward with his theory of gravity, sometimes referred to as the general theory of relativity.

The two parts of Einstein's work have one and the same intellectual origin, that of imposing Lorentz invariance on physics. Strictly speaking, relativity is not a theory by itself, but a requirement to be satisfied by physical theories.

For this and other reasons, many physicists regard the terms "special" and "general" theories of relativity as ghastly misnomers. Einstein himself wished later that he had used the term "invariant theory." He was particularly vexed by writers who seized upon the word "relativity" and associated it with other areas of human endeavor. For example, the novelist Lawrence Durrell claimed that he based the "four-decker form" of his masterwork, *The Alexandria Quartet,* on the relativity structure of time and space. In *Balthazar,* he proclaimed that relativity is "directly responsible for abstract painting, atonal music, and formless . . . literature." Why Durrell and others would not let their achievements in the arts speak for themselves is beyond me and many other theoretical physicists. One also encounters in ill-informed writing absurd statements asserting, for example, that Einstein proved that truth is relative. In fact, as we have seen, the whole point of Einstein's work is that different observers must perceive the *same* structure of physical reality and that an invariant truth *can* be extracted.

THE HAPPIEST THOUGHT AND THAT SINKING FEELING

By 1905, the same distaste for action at a distance that had led Faraday and Maxwell to formulate the electromagnetic field had also led physicists to describe gravity in terms of a field. They pictured a massive object, such as the earth, producing a gravitational field around it, in analogy to the electromagnetic case. Another massive object, be it a ball or the moon, feels this field and responds accordingly.

Aside from the new formulation in terms of fields, Newton's theory of gravity had survived intact for more than two hundred years, and indeed, it stubbornly resisted Einstein's efforts to make it Lorentz invariant. The crucial idea came in 1907. At that time, Einstein was working as a patent clerk in Bern. He was sitting in the patent office daydreaming when what he was later to call "the happiest thought in my life" suddenly occurred to him.

Before I explain what made Einstein so happy, I must remind the reader of a well-known fact about gravity. When Galileo (supposedly) dropped two iron balls with different masses from the Leaning Tower of Pisa and observed that they landed at the same time, he was trying to verify that objects of different masses fall at the same rate. The notion that a feather and an iron ball would fall at the same rate, in the absence of air resistance, seemed mind-boggling to Galileo's and Newton's contemporaries, but it was and is incontrovertibly true.

Starting with Galileo, many experimenters, notably Hungarian Roland Lorand, Baron Eötvös of Vásárosnamény in the late nineteenth century, and American Robert Dicke and Russian Vladimir Braginsky in recent years have verified this fact about gravity with ever-increasing accuracy. Newton incorporated this fact into his theory by simply supposing that the gravitational force on a particle is proportional to its mass. In Newtonian mechanics, the

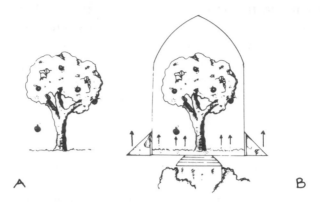

Figure 5.1. (A) An apple falls out of a tree. (B) A mad physicist plants an apple tree in a rocket, drives the rocket deep into space far away from any gravitational field, and then steps on the gas, accelerating the rocket at a constant rate. An observer floating outside the rocket might say that the ground inside the rocket is rushing up to meet the apple, but an observer inside the rocket will see the apple falling. Einstein asserted that no physical measurement can distinguish between the motion of the falling apple in A and in B.

acceleration of a particle of mass m acted upon by a force F is given by the well-known formula $F = ma$. The letter a denotes acceleration. Therefore, the mass drops out in determining the acceleration of a falling particle. In other words, Newton reduced the question of why objects fall at the same rate to the question of why gravity is proportional to mass. As Einstein set about constructing a relativistic theory of gravity, he insisted that the theory must account for this one essential fact, that all objects fall at the same rate.

Like many profound ideas in theoretical physics, Einstein's happy thought is marvelously simple. It is based on a common experience felt in the stomach while riding fast elevators. As the elevator accelerates upward, the elevator floor pushes our bodies up, but our stomachs, being loosely attached to our skeletal frames, cannot keep up. We sense our stomachs "sinking" momentarily. One might also argue that instead of our stomachs sinking toward the elevator floor, the elevator floor is "falling" up toward our stomachs. A similar phenomenon is experienced by passengers in rapidly accelerating cars and in this age of space travel, by astronauts.

Einstein imagined an elevatorlike box floating in space *far from any gravitational field*. Inside the box, various objects, say iron balls, also float in the total silence of space. Suppose the box starts accelerating at a constant rate. The iron balls continue to float, in happy ignorance of the fact that the "floor" of the box is rushing at them with ever-increasing speed. But to an observer sitting on the floor, it appears as if the iron balls are falling down toward the floor. What's more, the iron balls will hit the floor at precisely the same instant, in accordance with Galileo's observation.

Einstein was thus led to enunciate the principle of equivalence: In a small enough region of space, the physical effects of a gravitational field, as perceived by an observer, are indistinguishable from the physical effects reported by another observer accelerating at a constant rate in the absence of a gravitational field. In other words, acceleration can "fool" you into thinking that you are in a gravitational field. The rate of acceleration required depends, of course, on the strength of the gravitational field to be "mocked up."

Note that Einstein is *not* saying that apples fall because the earth is accelerating upward. If so, apples on the other side of the globe would "fall" upward. Einstein merely said that an apple's fall may be *equivalently described* by thinking of the apple tree accelerating upward. This is where the preceding paragraph's caveat, "in a small enough region," comes in. The equivalence principle applies only in a region over which the gravitational field is uniform in magnitude and direction. In our example, the region surrounding the apple tree obviously cannot be so large that the curvature of the earth comes into play. As we will see in Chapter 12, this local character of the equivalence principle is to have a far-reaching impact on contemporary theoretical thought.

Einstein's insight made physicists very happy. The equivalence principle offers a fabulously powerful, labor-saving method to advance our understanding of Nature. Suppose, for example, we want to know the laws of electromagnetism in the presence of a gravitational field, in order to study the behavior of photons in the vicinity of a black hole. It would seem that we would have to repeat the entire nineteenth-century experience, beginning with careful measurement of the influence of gravity on electromagnetic phenomena. The equivalence principle, fortunately, comes to the rescue. We merely have to work out what Maxwell's equations would look like to an observer accelerating at a constant rate.

In general, once we master a physical law in the absence of

gravity, be it the phenomenological law governing the flow of water or the more basic law governing the behavior of neutrinos, we can immediately find out what the law is in the presence of gravity by appealing to the equivalence principle.

GENERAL COVARIANCE

Einstein's original idea about the equivalence principle was inspired by the situation in a constant gravitational field, such as the one that affects us in everyday life. Most gravitational fields are not constant, however. The gravitational field of the earth decreases with distance from the center of the earth; to a good approximation, we feel a constant field in our daily lives merely because we don't significantly change our distance from the center of the earth. Following Einstein, we will now figure out how to apply the equivalence principle to a gravitational field varying in spacetime.

Einstein's strategy was exceedingly simple: Divide up spacetime into small regions, small enough so that within each region the gravitational field is constant. The situation is analogous to what geographers face in mapping the earth's curved surface. Geographers divide up the terrestrial surface into many small regions in such a way that the surface of each region is approximately flat to whatever accuracy is desired by the user of the map. Thus, for military use the region covered by each map must be rather small. (Geographers have other tricks, of course, such as using contour lines to indicate local topographical features, but these need not concern us here.)

To see how this strategy works in practice, suppose we are floating deep in space, far from any gravitational field. If we want to study a constant gravitational field, we know what to do: We hire a research assistant, put him in a rocketship, and accelerate it at a constant rate. He is to report his observations to us. But now suppose we want to study physics in the presence of a varying gravitational field, such as that surrounding two black holes orbiting about each other. See Figure 5.2A.

Thanks to Einstein's happy thought, we do not have to travel the cosmos in quest of an orbiting pair of black holes. Einstein instructs us simply to imagine dividing the spacetime around the pair of black holes into small regions such that, within each

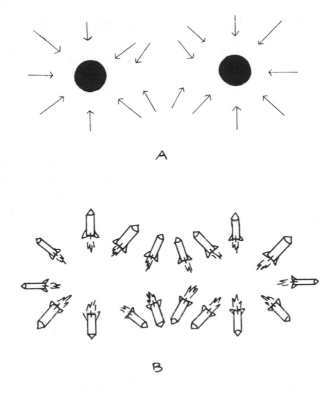

Figure 5.2. (A) The gravitational field around two black holes: Each arrow indicates the direction in which an object would fall if placed at the location of that arrow.

(B) According to the equivalence principle, we can study the physics around two black holes by finding a region deep in space far from any gravitational field and by accelerating a large number of rockets, each with a research assistant in it.

region, the gravitational field is effectively constant and uniform. Then we give our imagination free rein. In each region, we place a hypothetical rocketship. We then hire a lot of research assistants, put each of them in a ship, as indicated in Figure 5.2B, and accelerate each at a constant rate, depending on the strength of the gravitational field in that region. (The notion of dividing up time into small regions or segments is just a fancy way of saying that we have to carry out this experiment fast enough so that the gravitational field does not change effectively during the experiment.) We now merely have to read the reports of our research assistants to learn about physics in the gravitational field of an orbiting pair of black holes.

We have just carried out what is known as a thought experiment. We never had to leave home; all we had to do was work out how known physics would look to an observer accelerating at a constant rate. Neat trick, eh? The actual calculation involves a rather straightforward exercise in transforming coordinates.

Recall how, in the preceding chapters, we considered two observers in relative motion at constant velocity. Let (t', x', y', z') and (t, x, y, z) denote the spacetime coordinates assigned to a given event by the two observers. We learned that there is a collection of formulas, known as the Lorentz transformation, which tells us what t', x', y', and z' are in terms of t, x, y, and z. Here we consider two observers accelerating relative to each other at a constant rate. There ought to be, and there is, also a collection of formulas relating (t', x', y', z'), the coordinates used by one observer, to (t, x, y, z), the coordinates used by the other. Obviously, these formulas depend on the rate of acceleration. Remarkably enough, we do not need to know any further details about these formulas in order to go through the following development.

In our specific example, let (t', x', y', z') be the spacetime coordinates used by our research assistants, and (t, x, y, z) the ones we use. Since this is a thought experiment, we spare no expense. We make the regions minuscule and we hire a zillion research assistants. As we vary (t, x, y, z), or, in other words, as we move about from region to region, we are going from one research assistant to another, and thus the coordinates (t', x', y', z') vary just as (t, x, y, z) vary in a way that depends on the precise character of the gravitational field surrounding the given orbiting black hole pair.

A theory of gravity must deal with all possible gravitational fields, and so we are led to consider spacetime coordinates (t', x', y', z'), depending on (t, x, y, z) in all possible ways. A change of spacetime coordinates, going from (t, x, y, z) to (t', x', y', z'), with (t', x', y', z') depending on (t, x, y, z) in an arbitrary and general way—in other words, in any way we please—is known as a general coordinate transformation. In contrast, with a Lorentz transformation, the two sets of coordinates are related in a specific way.

What can we conclude from all this? Suppose we have to study physics in the presence of some arbitrary gravitational field. According to the preceding discussion, we may study physics in the absence of gravity, then simply perform a general coordinate transformation.

Einstein was thus led to demand that the laws of physics preserve their *structural* form under a general coordinate transformation. This fundamental requirement is known as the principle of general covariance.

The reader can easily imagine how general covariance would constrain the possible theory of the world. Let us see how this works in practice. Suppose, after years of thought and experimentation, we arrive at a law of physics expressed typically as an equation describing how various quantities change as (t, x, y, z) changes. But, another physicist may come along and simply say that he does not like the coordinates we are using to describe spacetime. He prefers his own choice, describing spacetime in terms of (t', x', y', z'), related to our (t, x, y, z) in *any way he likes*. And yet, when we reexpress our equation in terms of (t', x', y', z'), it would have to be structurally the same as our original equation. It must have the same structure. Most equations would not be able to pass this test! And they must be rejected. Thus, if we accept general covariance, we need consider only a restricted class of theories.

A SUBTLE DIFFERENCE

There is a subtle difference between Lorentz invariance and general covariance as symmetries. Lorentz invariance asserts that two observers in relative uniform motion perceive the same physical reality; it is a symmetry in the same way that rotational invariance is. General covariance does not make the obviously absurd statement that an accelerating observer would also see the same physical reality; it says that this observer can interpret the difference between the physical reality he experiences and the physical reality the nonaccelerating observer experiences as being due to a gravitational field. General covariance is a statement about the nature of gravity. The eminent American physicist Steve Weinberg has suggested that one refer to general covariance as a dynamical symmetry to underline the distinction. In this book, I follow customary usage among physicists and refer to general covariance simply as a symmetry.

OUR PERCEPTIONS UNDER COORDINATE TRANSFORMATIONS

The analogy between choosing coordinates to describe spacetime and choosing coordinates to describe the round earth on a flat page in an atlas is apt in some respects, but awfully misleading in others. In the standard Mercator map of the earth, the areas near the two poles are stretched out; indeed, many of us grew up thinking that there was an enormous continent named Greenland in the North Atlantic. The analog of the concept of coordinate invariance in cartography states the obvious, that the actual area of Greenland cannot possibly depend on how large it looks in an atlas. Similarly, physicists insist that physical reality cannot depend on the coordinates used.

In a striking recent work, West German historian Arno Peters has emphasized how the "eurocentric" Mercator projection has distorted our geopolitical perceptions. Consider the geopolitical division of the world into the "rich north" and the "poor south." Peters points out that since much of the "poor south" lies around the equator (an economic historian would add that this very fact accounts for the poverty of the "south"), the "poor south" looks much smaller than it really is in relation to the "rich north" in Mercator's map. Peters has published a new map of the world in which nations are represented with their true relative sizes. In the so-called Peters map, the world looks strikingly different. Interestingly, but not surprisingly, Peters writes that the publication of his map provoked a "vehement public discussion [in Europe], up to then unknown in the history of cartography."

THE THEORY OF GRAVITY

General covariance is a stringent requirement. Indeed, it was precisely because of this extreme stringency that Einstein was able to find the correct theory of gravity. In 1907, thanks to his happy thought, Einstein discovered how to describe physics in the presence of gravity, but only if physics is already known in the absence of gravity. But what about the physics that governs the dynamical behavior of the gravitational field itself?

The question stumped Einstein for years. No experiment was available to guide him. Because the gravitational force is so incredibly feeble, it is not feasible to perform direct experimental tests on the dynamics of the gravitational field, as distinguished from the dynamics of matter moving in a given gravitational field.

So how could Einstein proceed?

Einstein's own brainchild, the principle of general covariance, came galloping to the rescue. Historically, it took Einstein an agonizingly long time to understand general covariance, but once he understood it, he was able to write down the physics governing the gravitational field almost immediately.

Here is a crude analogy: Suppose an architect is told to guess the geometrical shape of a large hall. The architect cannot start unless she is given some "experimental input," perhaps in the form of several photographs giving partial views of the hall. But now suppose the architect is told that the shape is invariant under rotation by any multiples of 60° around its center. This is potent information indeed. The architect can immediately narrow the possibilities for the shape down to a hexagon, a twelve-sided–gon, an eighteen-sided gon, and so on. The simplest guess would be the hexagon. In physics, too, the imposition of a symmetry immediately narrows down the possibilities. An unspoken rule among physicists dictates that all things being equal, one goes for the simplest possibility—a rule that has worked remarkably well.

WARPED TIME AND SPACE

Perhaps no other aspect of Einstein's work has gripped the public imagination more than all the mysterious talk about curved time and space. Actually, the notion of curved spacetime follows directly from the equivalence principle.

Consider this fairly well known riddle. A hunter walks south for one mile, then turns due east and walks for another mile. Finally, he turns due north. After walking for yet another mile, he arrives back at his starting point and shoots a bear. What color is the bear?

The distance and angle given in the riddle immediately tell us that the earth must be curved. In general, if we know the shortest distance between *any* two points, we can work out precisely how the surface is curved. I am talking of the actual or intrinsic distance, of course, a distance that does not depend on which world map we are using. The actual distance between Timbuktu and Katmandu is the distance experienced by an air traveler flying the shortest possible route.

Similarly, in physics, the intrinsic distance between any two points in spacetime can only be the proper time experienced by a traveler going from one point to the other. According to the principle of equivalence, the physics seen by an observer in a gravitational field is identical to the physics seen by an accelerating observer in the absence of any gravitational field. We have already learned that proper time as experienced by the train passenger and as experienced by the stationmaster can be quite different. By extension, an accelerating observer would notice yet another proper time. Thus, the presence of a gravitational field must change the relative positions between various points in spacetime.

Suppose we are handed an airline timetable on which someone has changed all the flight times. By noticing that the flight time between Katmandu and Timbuktu is actually shorter than that between Katmandu and New Delhi, we could conclude that someone has warped the surface of the earth so that it is no longer a sphere. In the same way, Einstein was forced to conclude that a gravitational field warps spacetime.

THE BENDING OF LIGHT

The warping of spacetime has dramatic consequences. In Euclidean geometry, the shortest path between two points is a straight line. But for an arbitrarily curved space, the notion of a straight line can no longer be defined. One still can talk meaningfully of the shortest path, however. Just as a ship's navigator seeking to travel the shortest path on the curved ocean surface is compelled to plot a curved course in space, a photon traveling toward us from a distant star is forced to follow a curved path as it passes through the gravitational field of the sun.

In 1911, Einstein predicted that starlight grazing the sun during a total solar eclipse would appear to be bent. (For the effect to be large, one wants the photons to pass as close as possible to the sun. But then for the star to be visible next to the sun, one needs an eclipse to cut down the glare.) Because of the utter weakness of gravity, the predicted bending is only two thousandth of a degree. (As a theoretical physicist, I am impressed that observa-

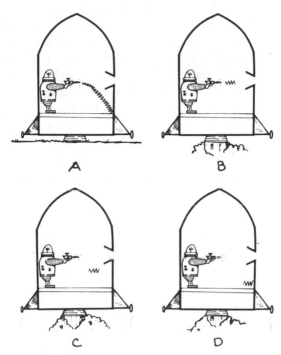

Figure 5.3. (A) A robot in a rocketship parked on an extraordinarily dense planet fires his laser gun, aiming it at a window. For the sake of clarity, the artist has exaggerated the strength of the gravitational field pulling the light beam down.

(B) The rocketship is now deep in space, far from any gravitational field, and accelerating at a constant rate. He aims his laser gun at the window and fires.

(C) The flash of light zings across the rocketship, oblivious that the floor of the rocketship is moving up at an ever-increasing rate.

(D) Instead of going out the window, the flash of light hits the foot of the opposite wall. As far as the robot can see, the trajectory of the light is the same as in A, thus verifying the equivalence principle. The robot cannot tell whether he is sitting still in a gravitational field or accelerating.

tional astronomers in the 1910s did not think this minuscule amount unmeasurable.)

It is instructive to see how the equivalence principle mandates that gravity must affect light, in spite of the fact that the particle of light, the photon, has no mass. Let us drug a research assistant into unconsciousness, put him in a rocketship, and send the ship deep into space. We accelerate the ship at a suitably constant rate, so that when the research assistant wakes up he is fooled into thinking that he is still on earth. (Remember, in these thought experiments we spare no expense: The ship's interior is decorated exactly the same as the assistant's living room!) We now tell the poor fellow to shine a beam of light at the wall. In the time that it takes the photons to get to the wall, the wall has moved upward (call the direction the ship is moving "up"), since the ship is accelerating. (See Figure 5.3.) Thus, the light beam will arrive at a spot below where it would have arrived had the ship been at rest. But the assistant, thinking that he is still in the earth's gravitational field, concludes that gravity is pulling the beam down!

This is how the equivalence principle works: You fool someone into thinking he is in a gravitational field. The physics he sees is then declared (by Einstein) to be the physics in a gravitational field. Got to keep that one in mind for next year's April Fools'!

CANONIZATION

They say it takes three generations to learn how to cut a diamond, a lifetime to learn how to make a watch and that only three people in the entire world ever fully comprehended Einstein's Theory of Relativity. But football coaches to a man are convinced that none of the above is comparable in complexity to playing quarterback in the NFL. I mean, watches don't mix up defenses on you, diamonds don't blitz and Einstein had all day to throw. $E = mc^2$ doesn't rotate coverages.
—J. Murray, sports columnist, Los Angeles Times Syndicate, November 4, 1984

While his cogitations as a patent clerk had set the physics world on its ears already, in 1911 Einstein was not yet a household word. Interestingly enough, as astronomers set out to test Einstein's bold prediction on the bending of light, historical accidents conspired to ensure Einstein's dramatic public canonization. First, torrential rains washed out an Argentinian eclipse expedition in 1912. Then, a well-financed German expedition went to the Crimea to prepare for an eclipse scheduled for August 21, 1914—but the guns of August boomed. An infuriated Einstein wrote to a friend that "only the intrigues of miserable people" prevented his ideas from being tested.

In fact, Einstein was lucky. In late 1915, he himself discovered that he had made a mistake. Einstein had neglected, of all things, the peculiar effects of warped space. The bending he had predicted was only half of the correct value.

In 1919, with the war over, two English expeditions traveled to Brazil and to Spanish Guinea. Their observations of the solar eclipse dramatically confirmed Einstein's theory. If history had been otherwise, Einstein would have been somewhat embarrassed.

Instead, the incomprehensible mystery of the theory and the dramatic understandability of the experiment captivated the war-weary world. The press leaped on a story unrelated to famine, Bolshevism, and reparations. The London *Times* broke the news with the headlines REVOLUTION IN SCIENCE/NEW THEORY OF THE UNIVERSE/NEWTONIAN IDEAS OVERTHROWN/MOMENTOUS PRONOUNCEMENT/SPACE WARPED. *The New York Times* reported that twelve, count them, twelve, wise men in the world understood the new theory. Einstein was an instant celebrity, courted by statesmen from around the world.

WARPED TIME

That spacetime may be warped is still often presented in popular expositions with a certain science-fiction aura, although it has been established since 1919. That gravity warps time was verified in 1960 in a terrestrial experiment performed by R. V. Pound and G. A. Rebka of Harvard University. They set out to show that at two different points in a gravitational field, at the top and at the bottom of a tower, for example, time passes at different rates.

Pound and Rebka did not have to look far for a suitable tower. As Pisa has its tower, so too the Harvard physics department. A simple calculation using the equivalence principle predicts that two clocks placed at the top and bottom of the tower attached to the Harvard physics building would differ by one second after 100 million years. To detect such a fabulously tiny deviation, Pound and Rebka had to stretch their ingenuity to the limit.

The experimenters used a photon as a clock. We all know that electromagnetic waves oscillate at definite frequencies, a fact enabling us to tune in our favorite programs on radio and television. Thus, a photon, oscillating at some definite number of cycles per second, can serve as a clock. Pound and Rebka beamed photons down the Harvard tower and meticulously measured their frequencies at the top and at the bottom of the tower. If time passes at different rates at the top and at the bottom, then the frequency of a photon as it travels down the tower, should change slightly, as it did. The experiment dramatically confirmed Einstein's theory. Years later, Pound would joke that he learned the true meaning of gravity while lugging the heavy experimental equipment up and down the tower.

BLACK HOLES

Il est donc possible que les plus grands corps ... de l'universe, soient ... invisibles.
—Pierre Simon, Marquis de Laplace

Ingenious experimenters have continued to test, and to uphold, Einstein's view on gravity. Unfortunately, terrestrial and local solar system experiments appear limited in detecting minute differences between Newton's and Einstein's theory. The two theories differ dramatically only in strong gravitational fields, such as those surrounding black holes.

Black holes have captured the public imagination. In fact, the idea of the black hole is neither new, nor particularly profound. As early as 1795, the Marquis de Laplace, then associated with the Bureau of Longitudes and the National Institute of France, remarked that even light may not move fast enough to escape from an extremely dense astronomical object. The dense object pulls the light back.

That light could not escape from a sufficiently dense object is an obvious point that everyone agrees upon; the real question is how such an object could have formed. The standard scenario envisages a massive star collapsing after it has burned up its nuclear fuel. In 1939, J. Robert Oppenheimer, G. Volkoff, and H. Snyder pointed out that a sufficiently massive star could not arrest its own collapse and eventually would reach the critical density imagined by Laplace. Here, Newton and Einstein differ crucially. In Newtonian physics, one is free to suppose that the matter in the star could become stiff enough to resist collapse. But a lot of energy is necessarily associated with the stiffness. (Think of the energy placed into a compressed spring waiting to be released.) According to Einstein, that energy, being equivalent to mass, would generate an additional gravitational field, which, in turn, would hasten the collapse. In Einstein's theory, one cannot avert, even in principle, the impending formation of a black hole.

THE MAN WHO WOULD NOT LISTEN

Over a distance of a few miles, the flat-earth theory and the round-earth theory differ negligibly. But as one travels over longer and longer distances, the differences between the two theories become more and more drastic until finally, when one has circumnavigated the globe, one can see that the two theories are totally different. Similarly, the difference between Newton's theory and Einstein's becomes total when we consider the universe as a whole. In particular, since in Einstein's theory space is curved just as the earth's surface is, we may be able to circumnavigate the universe just as we can the world. Hop in a spaceship, keep going straight ahead (that is, follow the shortest route), and eons later you may come back to the same point. In this case, the universe is finite, and said to be "closed," curved like the surface of a sphere. It is also possible that the universe is curved more like the surface of a saddle. (Imagine the saddle to extend out indefinitely.) In this case, the universe is infinite in extent, and a space traveler flying straight ahead could keep on going forever, without ever revisiting any of the places already visited. In this case, the universe is said to be "open."

Evidently, astronomers are not able to decide whether the universe is open or closed by direct observation. Like the ancient Greeks trying to decide whether the earth is flat or round, physicists and astronomers must combine direct observations with indirect physical reasoning, and make their best determination. For the reader's information, the available evidence at the moment suggests that the universe is open.

In February 1917, less than two years after proposing the theory of gravity, Einstein initiated an exciting area in physics: modern cosmology. After a million years of evolution, the human mind was finally ready to reach beyond the stars to understand the cosmos itself. Einstein realized that since the motion of heavenly bodies is governed by gravity, a complete theory of gravity should tell us about the dynamics of the entire universe.

Nowadays, astronomers tell us that the universe is filled with a uniform distribution of galaxies. In 1922, the Russian Aleksandr Friedmann solved Einstein's equations for a universe uniformly filled with matter and showed that the universe must be either expanding or contracting.

We can easily imagine how shocking this notion was. The universe has always been thought of as unchanging and eternal. Indeed, Einstein himself was so wedded to the idea of a static universe that he thought his equations for gravity incomplete. In his 1917 paper, he actually changed his equations to allow for a static solution. Supposedly, he later referred to this move as "one of the worst blunders" of his life.

Before the jury convicts Einstein for not listening to his own theory, I must come quickly to his defense by reminding the jury of the astronomers' rather limited knowledge in 1917. At the time, they had not even established that there were galaxies other than our own Milky Way. But progress came rapidly.

By 1929, American astronomer Edwin Hubble, who gave up a law practice to study astronomy, had established that galaxies are flying away from each other. By 1935, the expanding universe was an observational fact. In 1946, George Gamow proposed the Big Bang picture of creation.

The idea is simple. We imagine running a film of our expanding universe backward. The film now shows the galaxies flying toward each other. Eventually, all the particles in the universe are on top of each other. At that point, the film breaks. If we now run the film forward, we see all the particles of the universe rushing explosively outward from one point. The Big Bang picture has successfully accounted for some of the observed features of the universe.

There is a curious footnote to Einstein's blunder. In 1917, Einstein "stopped" the universe from expanding by adding a new term to his theory, the "cosmological constant." Since the universe does expand, the cosmological constant should not be present in the theory. But so far, no one has ever been able to produce a *theoretical* argument showing that this term should not be present. This difficulty, the "cosmological constant problem," ranks as one of the most profound unsolved problems in physics today.

SECRET LABYRINTHS

Einstein's blunder illustrates again that a great physical theory contains in its inner structure secrets undreamed of by its creator. Theories should lead the theorists, and not the other way around. Einstein's theory led us all the way from the sinking feeling in the stomachs of elevator riders to the expanding universe. This is in sharp contrast to phenomenological theories, constructed simply to "explain" a given phenomenon. Theorists craft such theories to fit the data, and get out as much as they put in. They lead their phenomenological theories, rather than the other way around. Such theories may be of great practical importance, but typically, they tell us little, if anything, about other phenomena, and I find them to be of no fundamental interest.

IT MUST BE

Einstein's theory of gravity exemplifies theoretical physics at its awesome best. The starting point of the theory is rooted in everyday experience, yet its consequences are magnificently counterintuitive. Given that objects fall at the same rate, the human intellect is able to erect a theory from which the dark secrets known hitherto only to the gods—such as the gravitational distortion of time, and the evolution of the universe—flow as natural, logical consequences.

Einstein's theory of gravity carries with it a sense of the inevitable. The notion that a particular theory is the only one possible was new to physics. For instance, Newton's pronouncement that the gravitational attraction decreases as the square of the distance between two bodies appears quite arbitrary from a purely logical point of view. Why doesn't the force decrease as the distance, or as the cube of the distance?

Newton would have regarded this question as unanswerable; he presents his law simply as a statement whose consequences accord with the real world. Altogether different, once

Figure 5.4.   Two compatriots who understood the necessity of art.

Einstein enunciated the requirement of general covariance, the theory of gravity is fixed.

Abraham Pais, the leading biographer of Einstein, has aptly remarked that if Einstein's theory of special relativity in its perfection is reminiscent of a Mozart composition, then his theory of gravity has the full force of a Beethoven opus. The last movement of Beethoven's Opus 135 carries the motto: *"Musz es sein? Es musz sein."* (Must it be? It must be.) Art in its perfection must be a necessity.

Perfect art must be unalterable. Does anyone dare, or more to the point, even want to rewrite Beethoven's Ninth? In theoretical physics, the construction is tighter yet. Physicists are less respectful of authority than musicologists, and subsequent generations of physicists have toyed with Einstein's theory of gravity with an eye toward improving it. But there is no way to modify the theory significantly without abandoning general covariance. Under the aesthetic rule of symmetry, one can embellish Einstein's theory, but not modify his conclusions.

6

Symmetry Dictates Design

A wise, older colleague once told me that he reads Tolstoy's *War and Peace* once every ten years and finds it a different book each time. The greatest literary masterworks are those with several levels of meaning; they speak to the reader according to the reader's own experience and sensibility. In the innocence of my teen-age years, I thought that *Death in Venice* was a murder mystery and was disappointed. Later, I was astonished to discover the wealth of symbolism in the work.

When I was in high school, I was not only interested in crime mysteries. One day, I came across a popular account of Einstein's theories. Like the typical layman, I was captivated by the outlandish and bizarre aspects of Dr. Einstein's universe. Later, in college, after I had mastered enough physics and mathematics to understand Einstein's work, I marveled at the mathematical subtleties involved, and I saw Einstein's strange conclusions as perfectly logical consequences of his theory. But as I learned more physics and started doing research, I finally realized the true intellectual legacy Einstein bequeathed to my generation of physicists: nothing less than a new way of doing physics.

A SCHEMA FOR FUNDAMENTAL PHYSICS

To appreciate Einstein's insight, let us review the schema followed in developing that quintessential nineteenth-century theory, the theory of electromagnetism.

By fooling around with frogs' legs and wires, physicists saw that Nature behaves in a certain pattern, and that Her behavior can be described by a set of equations. The equations, once written down, sing out a song, waiting patiently for someone with ears to

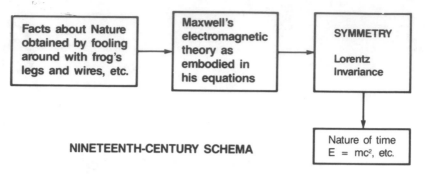

Figure 6.1. A large collection of experimental facts were summarized into equations which in turn revealed a symmetry in Nature's design. The symmetry, once seen, led to further empirically verifiable facts, such as the conversion of mass into energy. The connection of these facts with the facts of electromagnetism is far from obvious.

hear. Finally, a bright young fellow comes along and hears the equations saying they are Lorentz invariant. This fellow then realizes that the symmetry demands a revision of all of physics. (The schema for the development of electromagnetism and special relativity is illustrated in Figure 6.1.)

After Einstein worked out special relativity, it dawned on him and his contemporary Hermann Minkowski that the arrows in this schema may be reversible. Suppose that it was secretly revealed to us, in the dark of night, that the world is Lorentz invariant. Knowing this, can we deduce Maxwell's theory and hence, the facts of electromagnetism, without ever stepping inside a laboratory?

To a large extent, we can! The requirement of Lorentz invariance is a powerful constraint on Nature. Maxwell's equations are so intricately interrelated by this invariance that, given one of the equations, we can deduce the others.

Here is a flavor of the reasoning involved.

We are given a symmetry that relates space to time, the electric to the magnetic. Suppose that we know one of Maxwell's equations, say the one corresponding to Coulomb's law. You may recall that Coulomb's law describes how the electric field produced by a charge decreases as one moves away from the charge. In other words, we have an equation describing how the electric field varies in space. Under a Lorentz transformation, that equation is changed into an equation describing how an electric field varies in time, and how a magnetic field varies in space, corresponding to another one of Maxwell's equations precisely.

The essence of this symmetry argument can be stated succinctly. Since the symmetry unifies space and time into spacetime, and electric and magnetic fields into electromagnetic field, we cannot have an equation standing all alone describing the variation of the electric field in space. Indeed, as I remarked in Chapter 4, that equation can only be one piece of a unified equation describing the variation of the electromagnetic field in spacetime. In the architectural analogy, if the architect is told that a room possesses an exact hexagonal symmetry, and if she is shown a photograph of one wall, then she can obviously deduce the design of the entire room. In physics, the situation is mathematically more intricate, but the guiding idea is similar.

In Philip Roth's *The Ghostwriter,* one of the characters, a famous writer, tells another character that he always writes one sentence before lunch. After lunch, he turns the sentence around, and he spends his life turning sentences around and around in his head. In much the same way, theoretical physicists turn logical structures around and around in their heads. Thus, Einstein and Minkowski realized that one can turn the arrows around in Figure 6.1 and start with symmetry.

IN ONE FELL SWOOP

Einstein grasped the power of symmetry and put it to use in developing his theory of gravity. Instead of laboriously distilling this theory from a motley collection of experimental facts and then extracting a symmetry, he formulated a symmetry powerful enough to determine the theory. The schema he followed is illustrated in Figure 6.2.

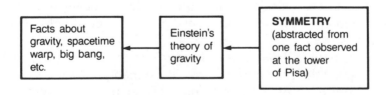

TWENTIETH-CENTURY SCHEMA

Figure 6.2.   The logical process followed by Einstein in discovering his theory of gravity. Contrast this process with the one followed in developing the theory of electromagnetism and special relativity (Figure 6.1).

Symmetry empowered Einstein to write down his theory of gravity in one fell swoop. To appreciate this, let us imagine what would happen if physicists followed the nineteenth-century schema in studying gravity, as some physicists tried to do. After years of carefully studying planetary orbits, astronomers would have noticed absolutely minute deviations of the orbits from the Newtonian prediction. To account for this, physicists would add a tiny correction to Newton's law of gravity. More careful study would reveal that this is still inadequate, and physicists then would be compelled to correct Newton's law by an even tinier amount. In practice, this program would quickly grind to a halt. But even if we imagine that physicists are able to determine as many correction terms as they like, it would take a stroke of mathematical genius to see that the corrections would all combine to produce a rather different theory. The theory in the intermediate stage would be a complicated mess. It is as if an architect had designed a square building, but the client really wanted a circular one. Each time the architect presented the drawings, the client would demand some small corrections but would refuse to tell the architect what he ultimately wants. The architect would keep on modifying her square design. Eventually, as the design looks rounder and rounder, she might realize that the client had wanted a circular design.

VOICES IN THE NIGHT

I regard Einstein's understanding of how symmetry dictates design as one of the truly profound insights in the history of physics. Fundamental physics is now conducted largely according to Einstein's schema rather than that of nineteenth-century physics. Physicists in search of the fundamental design begin with a symmetry, then check to see if its consequences accord with observation.

But how is a physicist to get to square one in playing Einstein's game? the reader might ask. Presumably, no one is going to come in the dark of the night and whisper to us the symmetries Nature has woven into Her tapestry. If an architect's client wants to have symmetrical designs, but won't tell the architect what symmetry he has in mind, how is the architect to find out?

Obviously, one can extract the symmetry from known experimental facts. That is what Einstein did. The difficult part is to decide on the one most relevant fact that allows formulation of a symmetry. Out of the many facts known about gravity, Einstein fastened on to the fact that objects fall at the same rate, regardless of mass. He did not use, for example, the fact that the gravitational attraction between two objects weakens as the square of the distance between them increases. This and all other known facts emerge as consequences of the symmetry imposed on gravity.

Another approach physicists are using with increased boldness as their discipline progresses is to listen to that half of the brain concerned with aesthetics. To read His mind, they search their own minds for that which constitutes symmetry and beauty. In the silence of the night, they listen for voices telling them about yet-undreamed-of symmetries.

Referring to the analogy used a bit earlier, we can imagine the architect trying hard to detect any hints regarding the symmetry that the client wants by poring over what was said by the client. This approach corresponds roughly to physicists trying to extract symmetry from observation. But the architect could also adopt a bolder approach by going ahead and devising the most harmonious design she can come up with. Then the architect can only hope that the client shares her aesthetic sense.

In Part Four, I will explain how both of these approaches have been adopted by physicists, to marvelous effects.

IN THE FOREST OF THE NIGHT

In his work on relativity, Einstein dealt with two interactions, electromagnetism, and gravity, which manifest themselves in the macroscopic world of everyday experience and about which we have built up a considerable amount of intuitive understanding. But even as Einstein worked, the old order was crumbling in the world of physics. The microscopic world of atoms and nuclei was found to dance to a different tune. The stately waltz of classical physics was replaced by the jitterbug of quantum physics. New interactions rule this strange microscopic world, interactions about which physicists had no intuitive feeling. There is more to the world than was thought by the late nineteenth-century physicists in their smug, deterministic complacency. Physicists have entered the forest of the night, where common sense is but a siren song leading them to deadly paradoxes.

Figure 6.3.   The Burning Tiger in the forest of the night.

In this darkness, the Burning Tiger, with his fearful symmetry, appears as a beacon of hope. Fundamental physicists have come to rely on the Tiger more and more. Today, symmetry considerations play the central role in the work of many fundamental physicists, myself included.

INTO THE
LIMELIGHT

7

Where the Action Is Not

In science, one tries to say what no one else has ever said before. In poetry, one tries to say what everyone else has already said, but better. This explains, in essence, why good poetry is as rare as good science.

It would appear that science and poetry are in extreme contrast to one another. However, some theoretical physicists, like poets, do devote their creative energies to saying what has already been said, but in a different way. Their work is often dismissed by more pragmatic physicists for essentially the same reason that poetry sometimes is dismissed. A body of physics is reformulated, but the new formulation does not advance our knowledge one whit. In the vast majority of cases, in poetry as in theoretical physics, the rude dismissal is perfectly justified. The new version is more convoluted and turgid than the old. But, once in a while, a poem, compact in structure, eloquent in cadence, manages to illuminate a theme more lucidly than ever before. In physics, too, formulations more in tune with the inner logic of Nature emerge from time to time. Perhaps the best example is the so-called action formulation, developed in the eighteenth century as an alternative to Newton's differential formulation of physics.

In Newton's formulation, one focuses on the particle at every instant in time. A force acting on the particle causes the particle's velocity to change according to Newton's law, $F = ma$. Thus, one knows the particle's velocity at the next instant, and, by extension, the particle's position. By repeating this procedure, one determines the position and velocity of the particle in the future. This, in short, is the standard formulation with which every beginning student of physics has to grapple. The formulation is called differential, since one focuses on differences in physical quantities from one instant to the next. The equations describing these changes are known as "equation of motion."

With the action formulation, on the other hand, one takes an overall view of the path followed by the particle and asks for the criterion the particle "used" in choosing that particular path rather than some other path.

For a long time, the action formulation was regarded as nothing more than an elegant alternative. Meanwhile, physics continued to be formulated in terms of differential equations of motion. However, my generation of theoretical physicists has finally embraced the action formulation and jilted the differential formulation. We had a change of heart largely because the action formulation makes our search for symmetry in the fundamental design much easier.

LIGHT IN A HURRY

When a swimmer stands in a pool of water, his legs look shorter. The same phenomenon can be observed by dipping a spoon into a glass of water. This phenomenon is easily explained by the bending of light as it traverses the interface between two transparent media—here, water and air. In Figure 7.1 a light ray goes from the swimmer's toes to point A on the water surface, bends, then goes to the observer's eye, E. The observer's brain, judging the direction from which the light ray comes, decides that

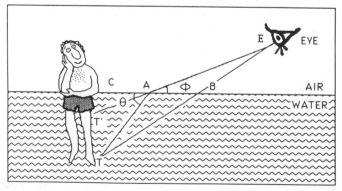

Figure 7.1.   Light in a hurry: In traveling from the swimmer's toe T to the observer's eye E, light "chooses" the path which enables it to get to its destination in the least amount of time. Since light moves faster in air than in water, the path TAE is chosen rather than the straight line path TBE. Then to the observer, the toe appears to be at T'.

it came from the point T'. Therefore, the swimmer's legs look shorter than normal.

To have a deeper understanding of why light bends in going from water to air, the mathematician Fermat (1601–1665) proposed, in the year of his death, a rather mysterious principle. Fermat's principle states that light chooses the path that allows it to arrive at its destination in the least amount of time.

In Figure 7.1 the straight-line path, *TBE*, is in fact shorter in distance than *TAE*, the path actually taken. But suppose that light moves more slowly in water than in air. Then by following path *TAE*, the photon traverses a shorter segment, *TA*, in water. The time saved makes up for the longer segment, *AE*, spent in air. What about path *TCE*? The segment in water, *TC*, is even shorter, but the segment in air, *CE*, is now longer. Clearly, there is an optimal path.

Motorists have to make the same kinds of decisions as the photon in Fermat's principle. In this age of high-speed expressways, the least-time route is often not the least distance. To travel from Paris to Venice, several routes are possible. A motorist may decide to cut across Switzerland, going through Zurich. Or, perhaps it is better to go around through the south of France, depending on the weather conditions.

(Speaking of motorists, I may mention another commonly observed optical illusion, also explained by light's propensity to hurry. Driving on a hot day, we can often see the "reflection" of distant cars in the road. Our brains, so easily deluded, conclude that the road ahead is wet. The phenomenon occurs because the layer of air near the road surface is hotter than the surrounding air, and the speed of light in air depends on the temperature.).

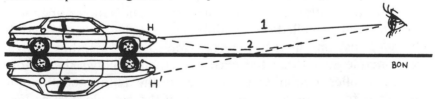

Figure 7.2. Summer mirage: A light ray leaving the hood *H* and headed downward encounters a layer of hot air near the road surface and bends upward. It ends up following Path 2 to the observer's eye. The observer's brain, judging the direction from which the light ray comes, concludes that it came from *H'*. Another light ray goes directly from *H* to the eye, following Path 1. This is repeated for light rays leaving every point on the car, causing a reflection of the car to be seen. The brain—what a marvellous organ—deduces that the road must be wet.

Fermat's principle so impressed his contemporaries that an analogous principle for mechanics was eagerly sought. In optics, the least-time principle frees us from having to memorize some not particularly illuminating formula relating the angles θ and φ in the drawing of the swimmer, Figure 7.1. Similarly, it was hoped that a compact principle would replace Newton's equations of motion.

The correct principle, known as the principle of least action, or action principle, was soon found by Pierre Louis Moreau de Maupertuis (1698–1759), Joseph Louis, Comte de Lagrange (1736–1813), and others.

WHERE THE ACTION IS NOT

The meaning of the action principle becomes clear by considering the prototypical process in which a particle starts at time t_A from point A and gets to point B at time t_B. In the action formulas, we consider not only all possible paths between A and B, but also all possible ways in which the particle can travel the path. Thus, for a specific path, the particle may go slowly at first, speed up for a while, slow to a crawl, then speed up again. Physicists refer to each particular way of traveling the path in the allotted time as a "history." In the action formulation, all possible histories must be considered. (Unlike photons, whose speed in a given medium is fixed, a massive particle can travel at variable speeds depending on the circumstances. Thus, in this respect the action principle is different from, and more general than, Fermat's principle. Indeed, physicists understood Fermat's principle later as a special case of the action principle.)

Let us continue with the statement of the action principle.

A number, called the "action," is assigned to each possible history. Thus, one history may be labeled 95.6, another 123.45. The principle states that the particle actually follows the history with the smallest action. Once we specify the action, the principle determines for us the actual trajectory of the particle, just as Fermat's principle determines the trajectory of light.

Physics can be formulated using the action principle. A given body of physics is mastered if we can find a formula that empowers us to determine the action for any history. For example, the action of a given history followed by a Newtonian particle is computed as follows: Subtract the particle's potential energy from its kinetic energy, then sum this quantity over the time period from t_A to t_B. (In Newtonian mechanics, the kinetic energy is simply the energy associated with the movement of the particle, while the potential energy is a kind of "stored" energy that is available for conversion into kinetic energy. For example, an object near the surface of the earth has potential energy because of the earth's gravitational pull. The higher the object is from the ground, the more potential energy it possesses. As the object falls, its potential energy is converted into kinetic energy. When we go downhill skiing, we pay the lift operator to provide us with lots of potential energy which we then convert into kinetic energy.) The computation of the action is similar to that done by an accountant determining the total profit of a business for any given production strategy. He subtracts the total cost of production from the gross income on a weekly basis and then sums this quantity over the fifty-two weeks in the fiscal year. The businessman naturally tries to maximize the total profit by following the most advantageous history.

A GREAT FALL

Let me illustrate how the action principle actually works with a falling particle. As indicated in Figure 7.3, Humpty Dumpty has to get from point A to point B in a specified time, while minimizing his action. Clearly, it does not pay for Mr. Dumpty not to fall straight down. To cover the larger distance of a curved path, Dumpty would have to move faster, thus increasing his kinetic energy and, hence, his action. Once Dumpty decides to fall straight down, he still faces a choice between an infinity of possible histories. To simplify things, Dumpty might begin by comparing two generically opposite strategic choices: He could go slowly at first, then speed up; or, he could go fast at first, then slow down. Recall that the action is equal to the quantity, kinetic energy minus potential energy, summed over the history. Since the potential energy increases with the distance from the ground, it clearly pays to spend more time high up, so that a larger potential energy could be subtracted off. Dumpty, therefore, starts slowly, then accelerates. With the help of elementary mathematics, one can show that the best strategy for Dumpty is to accelerate at a constant rate.

The reader may feel that, in this case, the action formulation actually is more convoluted than the differential formulation, and

Figure 7.3.   Mr. Dumpty has to decide which history minimizes his action in getting from *A* to *B*. The Path 2 does not minimize his action.

indeed it is. In the latter formulation, Dumpty's acceleration is determined immediately by Newton's law. However, as knowledge of physics progressed beyond Newtonian mechanics, the superiority of the action formulation became more apparent.

DIVINE GUIDANCE

I must emphasize that the action principle of mechanics says no more, and no less, than Newton's laws of motion. The action formulation, although more compact and aesthetically more appealing, is physically entirely equivalent to Newton's formulation.

The outlook, however, is quite different in the two formulations. In the action formulation, one takes a structural view, comparing different ways by which the particle could have gotten from here to there.

To the seventeenth- and eighteenth-century mind, the least-time and least-action principles provided comforting evidence of Divine guidance. A voice told each particle in the universe to follow the most advantageous path and history. Not surprisingly, the least-action principle has inspired a considerable amount of quasi-philosophical, quasi-theological writing, a body of writing which while intriguing, proves to be sterile ultimately. Nowadays, physicists generally adopt the conservative, pragmatic position that the least-action principle is simply a more compact way to formulate physics, and that the quasi-theological interpretation suggested by it is neither admissible nor relevant.

THE WORLD ON A COCKTAIL NAPKIN

The action principle turns out to be universally applicable in physics. All physical theories established since Newton may be formulated in terms of an action. The action formulation is also elegantly concise. For instance, Maxwell's eight electromagnetic equations are replaced by a simple action—by a formula enabling us to compute a single number for each possible history describing how the electromagnetic field changes. Similarly, the ten equations Einstein wrote down for his theory of gravity may be summarized elegantly in a simple action. (Einstein, and independently, the German mathematician David Hilbert, discovered the correct action shortly after Einstein arrived at his equations.) The point is, while the equations of motion may be complicated and numerous, the action is just a single formula.

The reader should understand that the entire physical world is described by one single action. As physicists master a new area of physics, such as electromagnetism, they add to the formula for the action of the world an extra piece describing that area of physics. Thus, at any stage in the development of physics, the action is a ragtag sum of disparate terms. Here is the term describing electromagnetism, there the one describing gravity, and so on. The ambition of fundamental physics is to unify these terms into an organic whole. While a mechanic tinkers with his engine, and an architect her design, a fundamental physicist tinkers with the action of the world. He replaces a term here, modifies another there.

Our search for physical understanding boils down to determining one formula. When physicists dream of writing down the entire theory of the physical universe on a cocktail napkin, they

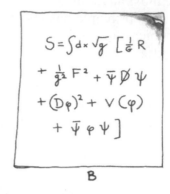

$$S = \int dx \sqrt{g} \left[\tfrac{1}{G} R \right.$$
$$+ \tfrac{1}{g^2} F^2 + \bar{\psi} \not{D} \psi$$
$$+ (D\varphi)^2 + V(\varphi)$$
$$\left. + \bar{\psi} \varphi \psi \right]$$

A

B

Figure 7.4. (A) Fundamental physicists dream of writing down the design of the universe on a piece of napkin. The action formulation allows an extraordinarily compact description.

(B) At present, most physicists believe the action looks something like what has been scrawled on this napkin. To understand what each symbol means, one would have to spend years in a reputable graduate school. However, you may notice all the plus signs right the way: This action consists of many pieces simply added together. For instance, the first term $\frac{1}{G}$ R represents gravity, while the second term $\frac{1}{g^2}$ F² represents the other three interactions. This indicates that physicists have not yet reached a completely unified description of Nature. As described in Chapter 16, physicists are struggling to find an even more compact action in which the six separate terms contained in this action will be tied together.

mean to write down the action of the universe. It would take a lot more room to write down all the equations of motion.

THE INVARIANT ACTION

Brevity well may be the soul of wit, but there is yet another, all-important reason to prefer the action formulation. Here, we come back to my central theme, symmetry. In discussing the concept of symmetry, I have taken great care to say that physical reality could appear different to different observers, but that the *structure* of physical reality must be the same. The action principle allows us to make precise the phrase "structure of physical reality."

As an illustration of this point, recall the discussion in Chapter 6 of how Coulomb's law changes under a Lorentz transformation. The mathematical equation for Coulomb's law has the form (electric field) = (function of charge). Under a Lorentz transformation, the quantities on the two sides of the equal sign both change, but in such a way that they remain equal. What looks like an electric field to the stationmaster is perceived by the train passenger as a combination of an electric and a magnetic field. Coulomb's law changes into Oersted's law.

In physicist's jargon, the equation is said to be covariant, rather than invariant. The two sides of the equation change in the same way, rather than remain unchanged. As a result, while the physical quantities involved change, the structural relationship between them does not. As a rough analogy, one can think of a marriage in which the two partners "grow" with the years. In those rare cases in which the husband and wife both grow in the same direction and at the same rate, the relationship between them would remain the same even though neither of them does. Unfortunately, psychologists tell us that most human relationships are not covariant in time (and most certainly not invariant).

In contrast to the equations of motion, the action is left invariant by a Lorentz transformation. The action remains unchanged. Indeed, to say that physics possesses a certain symmetry is to say that the action is invariant under the transformation associated with that symmetry. As a result, a history seen by different observers is labeled by the same number, 95.6, say, so there can be no dispute about which history is favored by the action principle. The action, in short, embodies the structure of physical reality.

To detect a symmetry in the fundamental design, one would have to check the covariance of each of the many equations of motion in the differential formulation. With the action formulation, on the other hand, one has the considerably easier task of checking the invariance of the action.

THINKING OF ACTION

With the advent of quantum mechanics, another basic reason for preferring the action formulation has emerged. It turns out that this formulation is naturally suited to describe quantum physics, as I will explain in Chapter 9.

For these and other reasons, in the field of fundamental physics, the action formula has elbowed equations of motion aside. In my own work, I have rarely, if ever, dealt with equations of motion and attendant concepts, such as force and acceleration.

Some physicists would like to believe that the Ultimate Designer thinks in terms of action.

8

The Lady and the Tyger

NATURE DOES NOT PUBLISH HER DESIGNS

Unlike an architect, Nature does not go around expounding on the wondrous symmetries of Her design. Instead, theoretical physicists must deduce them. Some symmetries, such as parity and rotational invariances, are intuitively obvious. We expect Nature to possess these symmetries, and we are shocked if She does not. Other symmetries, such as Lorentz invariance and general covariance, are more subtle and not grounded in our everyday perceptions. But, in any case, in order to find out if Nature employs a certain symmetry, we must compare the implications of the symmetry with observation.

The difficulty involved in working out the observable implications of a symmetry varies considerably, depending on the symmetry. The task is also complicated by the limited range of phenomena accessible to experimenters, and so the implications of certain postulated symmetries will perhaps never be subjected to direct verification.

We learned in the preceding chapter that a physical theory can be summarized by a quantity known as the action and that the symmetry of the theory is manifested in the invariance of the action under various transformations.

Einstein proclaimed that symmetry can dictate the form of the action. Physicists, however, are often faced with a situation in which they do not know all the symmetries involved, and the symmetries they do know are not restrictive enough. While they can narrow down the form of the action immensely, they may still be confronted with many possible actions. Tell an architect to impose bilateral symmetry and she can still construct a limitless variety of buildings.

Faced with this situation, physicists would have to examine

each "candidate" action in turn to determine its physical implications, a laborious process indeed. In extreme cases, it may take years, if not decades, to extract all the implications of an action that one can write down, literally with a few flicks of the wrist.

Now suppose someone comes along and asserts that, given a specific symmetry, one immediately can say what some of the implications are, regardless of the details of the action. Physicists would be overjoyed!

EINSTEIN IN SPIRITUAL ECSTASY

Early in this century, someone did come along: the mathematician Emmy Noether. Her profound observation remains the most general statement that physicists have about invariant actions. Einstein, writing about Noether in *The New York Times* after her death, said:

> Pure mathematics is, in its way, the poetry of logical ideas. One seeks the most general ideas of operation which will bring together in simple, logical and unified form the largest possible circle of formal relationships. In this effort toward logical beauty spiritual formulas are discovered necessary for the deeper penetration into the laws of nature.

Who is this Emmy Noether? And what is her "spiritual" discovery? Before I answer these questions, I have to explain the conservation laws of physics.

NO FREE LUNCH

The conservation laws of physics say that you get out what you put in, and no more. Nature says that there is no free lunch. Energy is conserved, and perpetual motion machines impossible.

Until the turn of the century, perpetual motion machines were quite the rage and were exhibited at fairs. Would-be inventors were obsessed with the idea that one could build a machine that would run forever with no fuel. Since a real-life machine is inevitably afflicted by friction, some energy must be supplied to keep the machine running. The machines that appeared to work were all eventually exposed as being of fraudulent construction, with hidden assistants, wires, and so forth.

In physics as in bookkeeping, the concept of conservation is important. The bookkeeper adds to the initial balance of an account all the payments into the account, subtracts all the payments out of the account, and checks that the sum equals the final balance. Nature does her own bookkeeping, with lightning speed, and has done so countless numbers of times since the world began. Experimental physicists, like independent auditors, have put Nature's books under the most minute scrutiny allowed by technology, and they have never found an error. The law of energy conservation has never been known to fail. Observe a collision of two billiard balls. Measure the speed of the balls before and after the collision. Compute the energy of movement (that is, the kinetic energy) corresponding to these speeds. While the energy of individual balls is changed completely by the collision, the total amount of energy is the same, before and after.

As our experimentalist improves his accuracy in measuring the speed of the billiard balls, he eventually finds a slight discrepancy. A tiny amount of kinetic energy is missing! Has Nature, like the computer thief lurking in contemporary banks, rounded off the last penny in each account for Her own profit? No, Nature has simply transferred the tiny amount to other accounts. With ever more delicate instruments, our indefatigable "auditor" now measures the energy carried off by the sound wave caused by the impact. He also detects that the billiard balls have become just a touch hotter, and even the table has become slightly warmer. When all forms of energy are included, the checkbook is balanced.

The concept of energy conservation is a great help to physicists in their computations. Let us give a simple example. Observe the mesmerizing swing of a pendulum. Knowing the gravitational force on the bob of the pendulum at any given instant, we can compute, by Newton's laws, how the speed of the bob is changing. Moving from one instant of time to the next, we determine the trajectory of the bob. However, it is much easier to recognize that, as the pendulum swings back and forth, energy is converted back and forth between kinetic energy and potential energy. Recall from Chapter 7 that the higher the bob is from the ground, the more potential energy it has. At the highest point, the pendulum is momentarily at rest and registers zero kinetic energy; its potential energy is at a maximum. At the lowest point, the kinetic energy is at a maximum, while the potential energy is at a minimum. The total energy, if we ignore small effects like air resistance, is conserved. At any given point on the pendulum's trajectory, we can determine the velocity from the kinetic energy simply by subtracting the potential energy at that point from the total energy. This is contrary to Newton's differential approach, in which one tries to follow the pendulum from one instant of time to the next. The conservation law approach is not only simpler, but in some sense, intellectually more satisfying.

Anyone who has been administered a dose of physics in high school knows that there are several other conservation laws. Momentum, for instance, is also conserved. In recent years, political writers covering American presidential elections also talk of momentum in a way that suggests some sort of conservation law. After a primary contest, one candidate is said to have the "Big Mo," which another candidate apparently has lost.

Conservation of energy and momentum is also of great practical importance to modern physics. At giant accelerators, physicists accelerate particles such as electrons and protons to enormously high energies and have them collide with each other in order to probe the secrets of Nature. These collisions send various particles flying off in different directions. In this fashion, physicists have discovered many hitherto unknown subnuclear particles, some of which live for only a short time. Their lifetime could be so short that even traveling at the speed of light, the particle leaves no detectable track before disintegrating into more stable, and more familiar, particles.

For example, an experimenter may detect an electron and a positron flying off at high speed. The experimenter proceeds on the working assumption that the electron and positron come from the same source, a disintegrating parent particle. Measuring the energy and momentum of the electron and of the positron, the experimenter then can determine the energy and momentum of the unseen, unknown parent particle by invoking conservation laws. Knowing the standard relation between energy, momentum, and mass of a particle, first presented by Newton and generalized by Einstein, the experimenter finally can figure out the mass of the unseen disintegrating particle.

The bookkeeping analogy we used earlier in this chapter is rather imperfect. That a checkbook should balance, although an accomplishment exasperatingly difficult to achieve at times, is completely obvious. We are simply verifying our ability to count correctly. That energy and momentum are conserved in all physical processes is more profound, and it tells us something about the inner design of Nature.

But what is energy? More precisely, given a set of equations governing how a physical system changes in time, we have to find a quantity that does not change. A priori, we would not know whether the kinetic energy of a freely moving particle is proportional to its velocity, its velocity squared, or its velocity cubed, and so on. More generally, given an action, how does one determine what is conserved?

Before Noether came along, physicists resorted to trial and error, juggling the given equations until they found a combination that did not change in time. Take the simplest case of two Newtonian particles interacting by a force that depends on the distance separating them. The two "particles" may be the earth and the sun, for example. As a first guess for the energy, a physicist might try the combination obtained as follows: For each of the two particles, multiply its mass by its velocity squared, then add the two quantities.

According to Newton, a particle's velocity changes at a rate given by the force acting on it divided by its mass. Knowing this, our physicist can easily calculate whether his trial quantity changes. It does. But if the physicist is clever enough, he might notice that if he adds to his combination a quantity that depends on the separation between the particles, then the total sum, lo and behold, does not change. He has found a conserved quantity, which he decides to call energy. Our physicist is lucky enough to have started with a correct first guess. If he had cubed the velocities instead of squaring them, or if he had failed to multiply by the particles' masses, no amount of juggling would have led him to a conserved quantity. Some readers might recall that high-school physics textbooks simply assert what the energy is, then verify a posteriori that it is indeed conserved. That is not how physics is done.

It would be an extreme nuisance if physicists had to adopt the trial-and-error approach, particularly when faced with the more abstract actions considered today. Furthermore, one does not know, a priori, how many conserved quantities the action contains.

THE LIFE AND TIMES OF EMMY NOETHER

Emmy Noether now comes to the rescue. A great mathematician, Amalie Emmy Noether (1882–1935) had to struggle for her right to be what she wanted to be. While women had been allowed into universities in France in 1861, England in 1878, and Italy in 1885, there was still enormous resistance at the turn of the century to women pursuing higher education in Germany. Typically, an eminent academic of the time had thundered that their admission to universities would amount to "a shameful display of moral weakness." Noether persisted and managed to earn a doctorate. But it was out of the question for her to hold any sort of academic position.

In 1915, eminent mathematician David Hilbert, whom we met already as the codiscoverer of the action for Einstein's theory of gravity, recognized Noether's ability and invited her to join him in Göttingen, then a leading German center of learning. Hilbert tried in vain to obtain for her the right to lecture, without pay. One can almost hear the outcry: "First, they want to study; now, they even want to lecture!" The request was officially rejected because of "unmet legal requirements." The right to lecture had been reserved for males under a rule passed in 1908. At the faculty meeting, the philologists and historians would not budge, and an exasperated Hilbert stormed out, shouting something like "We are a university, not a bathhouse!"

World War I did not do Germany much good but it did bring changes to German society. In 1918, the legal status of women there was improved. Following an oral examination conducted by the faculty, Emmy Noether was given the right to lecture. There was considerable grumbling by the Old Guard that soldiers who had defended the fatherland and who had suffered so much already would now have to listen to a woman.

SYMMETRY AND CONSERVATION

It was during Noether's faculty examination to establish that she was good enough to lecture without pay that she presented her famous result. She had been studying actions invariant under symmetry transformations. Clearly, these actions should have special properties. But which ones?

It is useful to distinguish here between continuous symmetries, such as rotation, and discrete symmetries, such as parity. As the name suggests, one can vary continuously the transformation corresponding to a continuous symmetry. In the case of rotation, one can continuously vary the angle of rotation. With parity, however, either there is a reflection, or there isn't.

Noether, in a flash of insight, realized that for every continuous symmetry in the action there results a conserved quantity. Symmetry and conservation, two concepts beloved by physicists, are in fact connected!

The connection is not only profound, but also, as I have stressed, immensely useful. The experimental observation of a conserved quantity tells us immediately that Nature has incorpo rated a continuous symmetry in Her design. Electric charge, for instance, has been known to be conserved since the late eighteenth century. After Noether's discovery, physicists were prompted to reexamine the theory of electromagnetism and to search for the symmetry responsible for charge conservation. In this way, a deeper understanding of a theory that has been around for almost a century was obtained. The symmetry was duly found and became known as "gauge symmetry." In later chapters, we will see that the notion of gauge symmetry proved to be the key that enabled physicists, literally, to unlock the universe.

Noether's insight helped physicists in a multitude of ways. As physicists began to explore the nuclear and later, the subnuclear world, they would have no idea what the action was, but they might notice that certain quantities were conserved. Noether's observation tells them that the action must have a corresponding symmetry. Physicists are now able to take at least a first guess on what the action might be. Later, we will see this strategy successfully applied. If physicists had earlier been like half-blind art critics trying to discern the symmetries in Nature's tapestry, Emmy Noether gave them sight.

Conversely, if we know which symmetry transformations leave a given action unchanged, we now know immediately how many conservation laws there ought to be. Recall the physicists flailing away by trial and error to find conserved quantities. No more trial and error! Emmy Noether figured out how to determine the conserved quantities.

The beauty of Noether's observation is that it does not depend on the details of the action. Thus, several different actions invariant under the same symmetry transformation would necessarily all have the same conservation laws. Physicists no longer have to work on each action in turn to find what the conservation laws are.

PLUS ÇA CHANGE

Conservation of energy and momentum had been known for centuries, but physicists never linked them explicitly with symmetries. In light of Emmy Noether's insight, it is instructive to ask what symmetries are responsible. Since energy and momentum are so basic, the corresponding symmetries must be absolutely universal in character. What can they be?

Using Noether's theorem, one finds that energy is conserved if the physical laws do not change with time. In more technical language, the condition is that the action is invariant under a shift (or a "translation," to use the correct term) in time. But that is exactly what we want of physical laws. We want physics to be the same yesterday, today, and tomorrow!

We can easily understand the condition for energy conservation by considering a simple example in which energy is *apparently* not conserved. Envision a playground swing. A parent gives a child in the swing a firm push. One can say that the laws of physics, as perceived by the child, change with time. The child "feels" that the force acting on the swing changes. Of course, energy apparently is not conserved, but only because we choose to focus on the movement of the swing. When we examine the larger system, consisting of the swing, parent, and earth, energy *is* conserved, of course.

What does Noether's theorem say about momentum conservation? It turns out that momentum is conserved if the action is invariant under translation in space. In plain English, momentum conservation follows if physics is the same here, there, and everywhere. Again, let me illustrate by a simple example. Suppose I roll a ball toward a hill. As the ball climbs the slope, it loses momentum. Momentum appears not to be conserved. Once again, we are focusing narrowly on the ball. The physical laws "experienced" by the ball indeed change in space, according to whether it is on the slope or not. In fact, as I roll the ball in one direction, I cause the entire earth to move off in the other, by virtue of my attachment to the earth by gravity and friction. As the ball climbs the slope, and slows down in the process, the movement of the earth in the other direction also slows. The total momentum of the entire system is conserved.

Another basic conservation law states that angular momentum is conserved, a fact manifested most elegantly in the art of an Olympic ice skater. As the skater tucks her arms in, angular momentum conservation requires her to spin faster. Noether's theorem reveals that angular momentum conservation follows from rotational invariance. The physical laws are the same, regardless of which direction the skater faces.

Conservations of energy, momentum, and angular momentum are among the first laws that one learns when studying physics. Together, they govern the movement of everything in the physical universe, from the collision of galaxies to the whirl of the electrons in atoms. For years, I did not question where these conservation laws came from; they seemed so basic that they demanded no explanation. Then, I heard about Noether's insight and I was profoundly impressed. The revelation that these basic conservation laws follow from the assumption that physics *is* the same yesterday, today, and tomorrow; here, there, and everywhere; east, west, north, and south, was for me, as Einstein put it, essentially spiritual.

This particular revelation ranks among the most memorable in my years of being a physicist. Having always been intrigued by the capacity of the human intellect to comprehend the universe, I only come across true insights, such as Noether's, rather infrequently. These insights delight, awe, and move me, because, as absolute truths, they are at once profound and simple. On the other hand, I, as a physicist, do not find the behavior of a nucleus or a crystal under this or that circumstance interesting in itself. In the phenomenological perception of the universe, again what is interesting now will be of little interest to a later generation. Already, the present generation of fundamental physicists regards the fantastic discoveries of particle physics twenty years ago as, to use Einstein's phrase, "this or that phenomenon." But the connection between symmetry and conservation will last forever.

9

Learning to Read the Great Book

> No one will be able to read the great book of the Universe if he does not understand its language, which is that of mathematics.
> —Galileo

THE MATHEMATICS OF SYMMETRY

The search for fundamental symmetries boils down to the study of transformations that do not change fundamental physical action—such transformations as reflection, rotation, the Lorentz transformation, and the like.

To describe the structural properties of transformations, mathematicians and physicists have developed a language known as "group theory." Here, I would like to develop some basic notions of group theory for later use. The next two sections are, of necessity, more mathematical. Indeed, they are the most mathematical sections of *Fearful Symmetry*. Fortunately, you do not have to master mathematical details in order to understand the rest of this book. What is important is that you have some understanding of the terms that I will use later. The salient points are summarized at the end of this discussion.

Actually, once you get over the initial fright and acquaint yourself with the jargon, you will find group theory natural and intuitive. Suppose you are told to study a bunch of transformations. What would you naturally want to know? Two types of information, basically. You want to know what is the net transformation if you apply two transformations in succession. This tells you how different transformations are related to one another. Second, you want to know how the transformations scramble various objects together. I will organize the discussion along these two natural lines of inquiry.

COMBINING TRANSFORMATIONS

Given two transformations, call them T_1 and T_2, it is natural to consider what would happen if we first perform T_1, then T_2. Physicists call this combined transformation $T_1 \times T_2$. In the beaver's lesson cited earlier, Lewis Carroll considered two transformations, boiling and gluing. If T_1 is boiling, and T_2 gluing, then the transformation $T_1 \times T_2$ would be the operation of boiling followed by gluing. In a more serious vein, we may consider rotations. For instance, T_1 may be a rotation of 17° around a certain axis, and T_2 21° around another. Then $T_1 \times T_2$ is the rotation obtained by performing the rotation T_1, and then the rotation T_2.

We can think of the operation of combining two transformations as a sort of multiplication. Indeed, ordinary multiplication of numbers may be thought of as a special example of this combining process. For instance, if an investor tripled his money in one year, we can say that he "transformed" every dollar he had into three dollars. Suppose he managed to increase his money fivefold the next year. The combined transformation, which we may call 3×5, turns every dollar into fifteen dollars.

In ordinary multiplication, the number 1 plays a special role; every number multiplied by 1 is equal to the number itself. The transformation that does nothing plays the corresponding role when we combine transformations. This transformation is called the "identity transformation," denoted by I. For example, for rotations, the identity transformation is just rotation through 0°, or no rotation at all.

The multiplication of transformations obeys the same rules as ordinary multiplication, except for one crucial difference: While $3 \times 5 = 5 \times 3$, the product $T_1 \times T_2$ is not necessarily the same as $T_2 \times T_1$. The order matters. This is not particularly surprising. Our daily lives are full of operations that must be performed in a definite order to be effective. I suppose that in Carroll's example, one gets different results depending on whether one boils or glues first. But, let us resume the full academic seriousness of this discussion and illustrate this point with rotations.

For definiteness, consider a person, a marine recruit in a boot camp, say, standing and facing north. When the drill sergeant shouts, "Rotate by 90° eastward around the vertical axis" (I presume that there is a more technical term for this maneuver in the military), our recruit turns to face east. Suppose the sergeant next

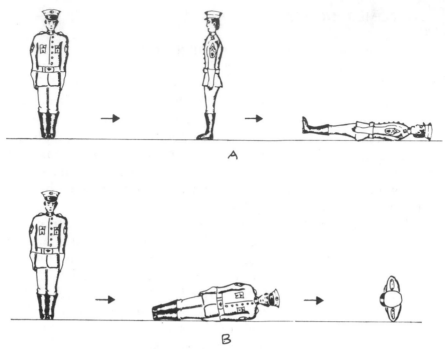

Figure 9.1.   (A) A marine recruit in bootcamp obeys two commands shouted by the drill sergeant.

(B) What would happen if the sergeant had reversed the order of his two commands?

shouts, "Rotate by 90° westward around the north-south axis." Our recruit ends up lying down on his back with his head pointing west, his feet pointing east. But what would happen if the sergeant reverses his two commands? You could easily verify that our recruit now ends up lying down on his left elbow, with his head pointing north. The order matters. For this reason, the study of rotations has been a *bête noire* for generations of physics students.

Fortunately for physicists, it turns out that mathematicians had already studied the multiplication of transformation in the nineteenth century, under the name "group theory." You have just learned that group theory involves a sort of advanced multiplication in which the order matters.

In contrast to physicists, who are preoccupied with the concrete, be it an actual physical object like an atom or a physical quantity like the action, mathematicians prefer to think of group theory in the abstract. As children, we went through a similar process of abstraction. We first learned that if there are three baskets, each containing five apples, then there are altogether fifteen apples. We then learned that as far as multiplication is concerned, it doesn't really matter whether the baskets each contain five apples, five oranges, or five kittens. According to my observations, children are able to abstract with remarkable ease, quickly learning to multiply without thinking of concrete objects. Similarly, mathematicians study group theory without referring to physical objects or situations.

To illustrate the preceding discussion on abstraction, I can use the two symmetries discussed in Chapter 3; namely, parity, and charge conjugation. For parity, the transformation is the reflection of our world into the mirror world; for charge conjugation, the transformation replaces particles with antiparticles. To a mathematician, the rules for multiplying transformation for these two symmetries are structurally identical: Two reflections in the mirror bring us back to our own world, and two charge conjugations bring a particle back to itself. The mathematician would say in either case that there is a transformation, T, such that $T \times T = I$. In other words, in either case, if you transform twice, you get back to where you started. She would concentrate on this relation, without caring one bit whether the physicist is considering parity, charge conjugation, or, for that matter, the interchange of yin and yang, in the same way that most of us can multiply without thinking about baskets of apples.

After all this, I am finally ready to define a group. A group is simply a bunch of transformations that can be multiplied together. If someone wants to describe a group to us, he has to tell us which transformations are contained in the group and to instruct us on how to multiply the transformations together. In the same way that ordinary multiplication is completely specified by the multiplication table that children are taught to memorize, a group is specified by its transformation and by a multiplication table. For example, the simplest group consists of two transformations, I and T. The multiplication table contains only four entries: $I \times I = I$, $I \times T = T$, $T \times I = T$, and $T \times T = I$.

Indeed, the first three entries just amount to the definition of I as the identity. It couldn't be simpler. This group, known as $Z(2)$, is relevant to physicists studying parity or charge conjugation.

As another example, the group named $SO(3)$ contains as elements all possible rotations in three-dimensional space. The multiplication rules are just those determined by performing two rotations, one after the other.

I am reminded of the story of a visitor to a joke-tellers' convention. One comedian would shout out "C-46!" and the other comedians would laugh appreciatively. Someone else would stand up and shout out "S-5!" and everyone would laugh. The puzzled visitor asked what was going on, and his friend explained: "All possible jokes, not counting minor variations of course, have been classified and numbered, and we all know them by heart." Similarly, all groups have been classified and numbered by mathematicians. When a physicist comes to my office, she might mutter $SO(3)$ or $E(6)$, and I would nod appreciatively. The physicist is telling me her guess on which group Nature uses in Her design.

Incidentally, I should give the punch line to the story. Finally, a comedian got up and shouted "G-6!" and everyone really cracked up. The visitor asked why this particular joke was so extraordinarily funny, and his friend replied, "Oh, that is Joe Schmo; he is so dumb that he doesn't know there is no such thing as G-6!" Similarly, if I were to mention $G(6)$ in a seminar, my colleagues would raise their eyebrows in surprise! Anyhow, all groups have been classified and named.

Fine, but what does this have to do with physics? As explained earlier, physicists are interested in transformations that do not change the action. Such transformations are called symmetry transformations. Now, if T_1 is a symmetry transformation, and if T_2 is a symmetry transformation, then $T_1 \times T_2$ is also a symmetry transformation. This statement is true, by definition. If neither T_1 nor T_2 changes the action, then, by performing T_1, then T_2, we obviously will not change the action. In other words, symmetry transformations form a group. Thus, physicists studying symmetries are led naturally to books on group theory. The group $Z(2)$ is, of course, so trivially simple that one hardly needs a course in higher mathematics to see its structure completely. But when physicists encounter more complicated groups, they are grateful that mathematicians have worked them out already.

Our example, that mathematicians can study $Z(2)$ abstractly without referring to parity or charge conjugation, though rather trivial, underlines the important point that various possible group structures relevant to physical theories, past, present, and not yet dreamed of, have already all been studied by mathematicians. Mathematics does not have to wait for physics.

REPRESENTATIONS

As suggested in the introduction to this chapter, we want to study next how the transformations in a given group scramble various objects together. The objects that are scrambled are said to furnish a representation of the group.

A representation of a group, roughly speaking, is a model of the group, much like an architectural model of a building. We expect the model to represent the structural arrangement of the actual building. The emphasis is on the *structural*. For instance, the relative sizes of two wings must be exactly the same in the model as in the actual building, but the color of the cardboard may be quite different from the color of the stone actually used.

To develop the notion of representing a group, let us, for the sake of definiteness, focus on $SO(3)$, the group of rotations in three-dimensional space.

Indicate the three directions in space by three arrows of some specified length, one pointing east, a second pointing north, the third pointing up. (See Figure 9.2.) For ease of speaking, let me label these three arrows $\vec{x}$, $\vec{y}$, and $\vec{z}$, respectively. We can indicate any other direction by writing something like $a\vec{x} + b\vec{y} + c\vec{z}$, with a, b, and c denoting three numbers. The three numbers may be thought of as constituting an instruction to a robot: For every a centimeters the robot moves eastward, it is to move northward by b centimeters, and to levitate upward by c centimeters. The direction in which the robot moves is the direction indicated by $a\vec{x} + b\vec{y} + c\vec{z}$. Thus, for example, the arrow $\vec{x} - \vec{y}$ points southeast, the arrow $\vec{x} - \vec{y} + 2\vec{z}$ points southeast and upward at an angle of about 55° from the horizontal. An arrow of the form $a\vec{x} + b\vec{y} + c\vec{z}$ is known as a linear combination of the three arrows $\vec{x}$, $\vec{y}$, and $\vec{z}$.

Now that we have learned how to specify directions, we are ready to go on to discuss rotations. We may describe a rotation by specifying the arrows that the three basic arrows, $\vec{x}$, $\vec{y}$, and $\vec{z}$, are rotated into. In other words, a rotation transforms each of the three arrows $\vec{x}$, $\vec{y}$, and $\vec{z}$, into a linear combination of the three.

Nothing profound or complicated is being said here; on the contrary, I am just expressing, in a precise way, the notion that a rotation scrambles the three directions of space.

In this example, a rotation is represented by its effect on the three arrows. The three arrows furnish a representation of $SO(3)$.

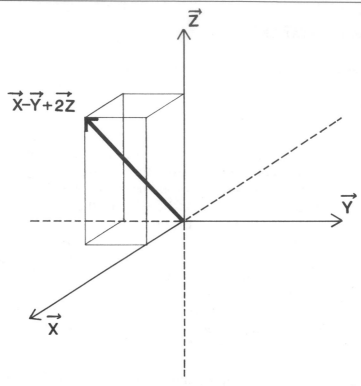

Figure 9.2. Adding and subtracting arrows: The three arrows $\vec{x}$, $\vec{y}$, and $\vec{z}$ point out the three directions east, north, and up. To determine the direction pointed out by the arrow indicated by the linear combination $\vec{x} - \vec{y} + 2\vec{z}$, travel east by one unit, travel north by minus one unit (i.e., travel south by one unit), and travel upward by two units. The direction you have traveled is the direction in question.

Since this representation exists practically by definition of $SO(3)$, it is called the "defining representation," or, sometimes the "fundamental representation."

You may feel that by this long discourse I have achieved little more than a restatement of the obvious: A rotation is defined by its effect on the three arrows. But here comes the remarkable point of studying representations. Using the defining representation, we can construct ever larger representations.

To do this, we, like the children we once were, throw away baskets, apples, oranges, kittens, and arrows. Instead, we think of the defining representation as furnished by three abstract "entities." After a rotation, each of these entities transforms into a linear combination of the three. To keep track of which entity is which, we have to give them names such as Huey, Louey, and Duey, or red, yellow, and blue. We may indicate these entities on the printed page as Ⓡ, Ⓨ, and Ⓑ. (The reader may think of these as three entities colored red, yellow, and blue, if that proves helpful.) Suppose we have three other entities that also transform as the defining representation. To distinguish between these entities and others, we write Ⓡ̲, Ⓨ̲, and Ⓑ̲. Purely for the ease of speaking, we refer to these two types of entities as "round" and "square," respectively.

Now we are ready to construct a larger representation by "gluing" these two copies of the defining representation together. We glue a round entity and a square entity together. In this way, we can form nine new entities, namely: Ⓡ [R], Ⓡ [Y], Ⓡ [B], Ⓨ [R], Ⓨ [Y], Ⓨ [B], Ⓑ [R], Ⓑ [Y], and Ⓑ [B]. Notice that we distinguish Ⓡ [Y], a red round entity glued to a yellow square entity, from Ⓨ [R], a yellow round entity glued to a red square entity. After a rotation, each of these nine entities obviously transforms into a linear combination of the nine entities.

It would appear that we have constructed a representation containing nine entities. But, wait! Here we have nine entities which are scrambled into each other by rotations. However, logically, that does not necessarily mean that any given entity can be transformed into *each* of the eight other entities.

Let me give a somewhat whimsical analogy. After a casual reading of fairy tales, an extraterrestrial might form the impression that the four objects, frog, prince, pumpkin, and carriage, may be transformed into each other. But, a more careful reading reveals that the four objects divided up into two separate pairs. The frog and the prince can be transformed into each other, but not into the pumpkin and the carriage. Here, by forming suitable combinations, we can divide up the nine entities into three separate clans: One clan contains five entities, another contains three entities, the third one entity. The nine entities can be divided up in the following sense. After any rotation, the five entities belonging to one clan transform only among themselves. In other words, the five entities transform into linear combinations of each other. They furnish a representation with five entities. Similarly, the three entities in one clan furnish a representation with three entities, and the single entity furnishes a representation with, well, one entity.

The situation is reminiscent of that at a Scottish village fair.

If we ask all those who are related to each other to stand together, the population splits up into clans. The analogy, admittedly, is imperfect, since the notion of transformation is missing.

Why these nine entities may be split up into separate clans is fairly easy to understand. Indeed, from a logical point of view, one may well ask why not. There is no reason to expect each of the nine obtained by gluing to be transformable into each of the other entities. The interested reader will find an explanation in the appendix to this chapter, page 285.

Instead of saying "a representation with five entities," mathematicians say "a five-dimensional representation." The use of the term "dimension" here is potentially confusing. We are discussing the representations of $SO(3)$, the group of rotations in three-dimensional space. This group, $SO(3)$, has a one-dimensional representation, a three-dimensional representation, a five-dimensional one, and for that matter, also one that is seventeen-dimensional. Thus, mathematicians use "dimensional" in referring to space and representations. That the group of rotations in three-dimensional space can scramble five or seventeen entities into each other is the sort of fact that a mathematician, but probably neither you nor I, would have thought of.

To summarize the preceding paragraphs, we say that the nine-dimensional representation we constructed by gluing two three-dimensional representations together splits up into a three-dimensional representation, a five-dimensional representation, and a one-dimensional representation. This fact is indicated by the equation $3 \otimes 3 = 1 \oplus 3 \oplus 5$. A representation is indicated simply by a number corresponding to its dimension. The act of gluing together is indicated by the $\otimes$ sign. (Notice that since entities cannot disappear into thin air, the "accounting" equation $3 \times 3 = 1 + 3 + 5$, obtained by omitting the circles from the equation $3 \otimes 3 = 1 \oplus 3 \oplus 5$, must also be true.)

By gluing two defining representations together, we encounter another representation, the five-dimensional representation. In it, rotations are represented by their effects of scrambling five entities together. By repeatedly gluing representations together, mathematicians generate all the representations of a given group. Having learned how to do $3 \otimes 3$, we can now go on and learn how to do $3 \otimes 5, 5 \otimes 5$, and so on. Thus, $3 \otimes 5 = 3 \oplus 5 \oplus 7, 5 \otimes 5 = 1 \oplus 3 \oplus 5 \oplus 7 \oplus 9$, and so on. (Here we encounter a seven-dimensional representation and a nine-dimensional representation.)

Some people actually pay good tuition to learn the rules for gluing representations together. For us, the important lesson to learn is not these detailed rules, but the fact that the group determines which representations are allowed. For example, $SO(3)$ has three- and five-dimensional representations, but not a four-dimensional one. It is not up to physicists to decide, on a whim, to construct a four-dimensional representation of $SO(3)$.

Our discussion for $SO(3)$ can be taken over for groups in general. When we come to grand unification later on, we will see that some physicists have proposed that the ultimate design of the world is based on $SO(10)$, the rotation group in ten-dimensional space. We can start with the ten-dimensional defining representation, and glue two such representations together, as before. In this case, it turns out that $10 \otimes 10 = 1 \oplus 45 \oplus 54$.

GROUP THEORY REDUX

Let me summarize the most significant points made in the last two sections.

1. The multiplication of symmetry transformations is not a capricious invention of the physicist; rather, the operation naturally suggests itself.

2. The multiplication structure of a group can be represented by the transformation of a number of entities. The number of entities involved is called the "dimension" of the representation.

3. The dimensions of possible representations are fixed, engraved in stone by the structure of the group. For example, the group $SO(10)$ has a 45-dimensional representation, but not a 44-dimensional or a 46-dimensional representation.

4. We can glue two representations together to obtain other representations.

WHICH MATH BOOK DID HE READ?

As physicists probe deeper into Nature, various integers start to appear. For example, we will see in a later chapter that the proton has seven "cousins." It turns out that the proton and its cousins furnish an eight-dimensional representation of a symmetry group. More traditional mathematics, such as calculus, is totally incapable of explaining the occurrence of special integers. Within our present mathematical framework, only group theory can explain why a certain number appears, and not another.

The occurrence of various integers gave physicists the first hint that Nature uses group theory and hence, symmetry considerations, in constructing Her design. One of the ultimate tasks of physics is to determine which group Nature chose.

The layman generally supposes that theoretical physicists use extremely complicated mathematics. Cartoons showing scientists often depict them in front of a blackboard covered with lengthy formulas. While that picture may accurately describe physicists in certain subfields, studying very complicated phenomena, anyone eavesdropping on two fundamental physicists at work is more apt to hear a heated exchange on such bits of wisdom as $10 \otimes 10 = 1 \oplus 45 \oplus 54$.

Toward the end of the last century, many physicists felt that the mathematical description of physics was getting ever more complicated. Instead, the mathematics involved has become ever more abstract, rather than more complicated. The mind of God appears to be abstract but not complicated. He also appears to like group theory.

10

Symmetry Triumphs

A STAR WAS BORN

Around the turn of the century, physicists discovered disturbing evidence that classical physics fails in the microscopic world of atoms. They eventually realized that classical physics, rooted firmly in our everyday intuition, is but an approximation of an underlying quantum physics. As the drama of physics moved from the classical to the quantum act, symmetry, already made a star by Einstein, was thrust into the limelight more than ever. First, I will introduce the reader to the mysteries of the quantum, then I will explain how the quantum cast symmetry into the leading role.

WORLD STABILITY

Let us follow one of the many strands in the historical development of quantum physics.

By 1911, it was established that an atom can be pictured as a miniature version of the solar system, with a number of electrons orbiting around an atomic nucleus. But electrons are electrically charged, and, according to Maxwell's theory of electromagnetism, a charge radiates electromagnetic waves when its motion changes. For instance, the electrons crashing through the filament of a light bulb emit electromagnetic radiation in the form of light. Maxwell's theory enables us to calculate the rate at which a moving charge radiates electromagnetic energy.

The orbiting electrons in an atom are changing their directions of motion constantly, and so, according to Maxwell, they should quickly lose their energy of motion by emitting electromagnetic radiation and spiral in toward the nucleus. Thus, according to classical physics, atoms should collapse in a very short time.

But in fact, atoms are quite stable—indeed, the very existence of the world depends on this fact.

The crisis brought on by the discovery of atoms was finally resolved by the Danish physicist Niels Bohr. Departing totally from established physics, in 1917 Bohr asserted boldly that an electron in an atom can occupy only certain orbits and not others. In classical physics, the orbit occupied by an electron depends continuously on its energy. If one makes the electron energy a bit smaller, it simply moves to a slightly smaller orbit. In Bohr's view, however, the electron can only have those energies associated with the orbits it is allowed to occupy. The energy of an electron in an atom is said to be "quantized."

Heeding Bohr's decree, the electron can no longer lose its energy continuously. Instead, it has to move to an orbit of lower energy by a "quantum leap." In Bohr's picture, the collapse of the world is averted simply because the electron has nowhere else to leap to when it reaches the orbit of lowest energy.

Bohr's contemporaries found it extremely difficult to swallow this picture, but they had no real choice in the face of the overwhelming experimental fact that the world has been around for quite a while. Yet, there were many puzzling questions. For example, if the electron can only occupy certain orbits, where can it be during the quantum leap from one orbit to another? Eventually, physicists recognized that such questions necessarily involve the classical and intuitive ideas of continuous motion, and they agreed not to ask.

As the electron leaps from one orbit to another, it emits a burst of electromagnetic radiation with an energy equal to the difference in energies of the two orbits, as required by energy conservation. Thus, the photons in the emitted radiation can only have certain definite energies. This is in startling contrast to the classical picture, in which the emitted radiation is expected to have a continuous spectrum of energy. In fact, experiments confirmed that the emitted radiation can only have certain definite energies.

THE FRENCH NEW WAVE

How can one possibly understand this puzzling quantization of energy?

In 1913, Prince Louis de Broglie of France made a brilliant

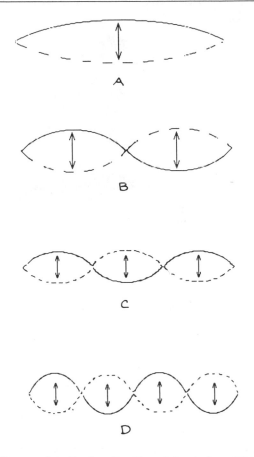

The wavelength of a vibrating string is "quantized"

Figure 10.1. The wavelength of a vibrating violin string is "quantized" merely because the string is tied down at two ends. If D is the distance from the bridge to the peg of the violin, the wavelength can be $2D$, D, $2D/3$, $D/2$ (as indicated in A, B, C, D, respectively) and so on, but it cannot be $1.76D$, for example.

suggestion. To understand the prince's idea, we will discuss for a moment the physics of music. When a violin is played, its strings vibrate, as illustrated in Figure 10.1, and produce music. Clearly, the possible wavelength of the vibration, defined as the distance from one crest of the wave to another, is determined by the distance, D, between the two points where the string is tied down. We see from the illustration that the wavelength can only be $2D$, D, $\frac{2}{3}D$, $\frac{1}{2}D$, and so on. The fact that only certain wavelengths are possible is, of course, why musical instruments produce definite tones. One could say that the wavelengths are quantized.

Figure 10.2. A French prince watching the electron wave going around an atomic nucleus: The wavelength is quantized because the electron wave has to catch its tail after going around. (Drawing adapted from *The Little Prince* by Antoine de Saint-Exupéry, copyright 1943, 1971 by Harcourt Brace Jovanovich, Inc. Reproduced by permission of the publisher.)

It occurred to de Broglie that this purely classical phenomenon of quantized wavelength may be relevant to energy quantization in the atom. He imagined that the electron is actually a wave propagating around the nucleus. As we can see from Figure 10.2, the wavelength of the electron wave can only take on certain values if the wave, after running full circle, is to catch its tail. This simple idea gave birth to quantum physics. At that time, Max Planck, Albert Einstein, and other physicists had already established that the energy and momentum of a photon is determined by the wavelength of the associated electromagnetic wave. Applying this result to the electron, de Broglie showed that his idea produces precisely Bohr's rule of energy quantization.

But physicists were perplexed by the nature of the prince's wave. De Broglie proposed what he called a "pilot wave," a sort of guardian angel guiding the electron along. Erwin Schrödinger, the Austrian physicist who formulated the equation describing the motion of de Broglie's wave, thought that the electron was literally stretched out, and made to wave, in the sense that a drop of water, if big enough, could fill a circular pipe and be made to wave. However, it was the German physicist Max Born who gave the interpretation most in accord with experiments. He suggested that the wave specifies the probability that the electron would be found in a particular place. The electron would most likely be found where the amplitude of the wave is the largest, namely at the crests and valleys of the wave. Yet the electron is still to be pictured as a point object, and not a spread-out fluid as Schrödinger thought.

This revolutionary suggestion signals the end of absolute determinism in physics. Probability now controls physics at the most fundamental level, causing no less a physicist than Einstein to moan, "The Good Lord does not play dice." Indeed, Einstein stubbornly refused to believe in quantum physics in spite of the evidence, and in spite of the fact that he himself was one of its chief architects.

Figure 10.3.   An artist's conception of God playing dice (after William Blake).

TO GET FROM HERE TO THERE

To underscore the probabilistic character of the quantum world, let us go back to the motorist traveling from Paris to Venice. Suppose he arrived in Venice and declared that he followed the route that got him there in the shortest time possible. Let us assume that we have the same perfect knowledge of road conditions and driving times that he has. Then we can determine exactly which route he took. This situation represents classical physics: The least action principle tells us precisely which path a particle follows to get from point A to point B.

Now, let us imagine that our motorist obeys the laws of the quantum. The situation changes drastically. When the motorist arrives in Venice, he can no longer declare that he came by the fastest route. He can only tell us that the probability is such and such percent that he came via Munich, and such and such percent that he came via Marseilles. Similarly, if Humpty Dumpty obeys quantum dynamics, then Mr. Dumpty, in keeping his fateful rendezvous with the ground, would not necessarily follow the path of least action. There is some probability that he might follow a path that would shock our everyday intuition, such as starting out fast and slowing down as he approached the ground. He might even get to the ground, not by falling straight down, but by following a curved path. Unlike classical physics, quantum physics can only tell us the probability that an object has followed a specific path. Of course, the probability borders on certainty that macroscopic objects follow the paths determined by classical physics.

Physicists believe in the probabilistic interpretation of quantum physics because a huge number of experiments have confirmed it. In an actual experiment, an electron is shot out of an electron gun, a device in which electrons are made to jump off a heated piece of wire. (A simple version of this can be found at the back of every television set. As the reader may know, the electrons in a television set are made to hit a screen, which, in turn, produces light when hit.) The experimenter has set up an array of electron detectors some distance away; they produce a signal when hit by an electron. He then places a screen with two holes between the electron gun and the detectors. This device is introduced merely to simplify the discussion. We can now focus on asking which hole the electron has gone through.

As indicated in Figure 10.4, detector number 5 has just sig-

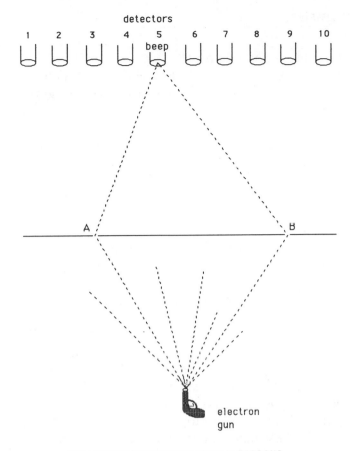

BALLISTICS EXPERIMENT WITH ELECTRONS

Figure 10.4.   A ballistics experiment with electrons: A screen with two slits, *A* and *B,* separate an electron gun from a bank of electron detectors. Detector number 5 has just signaled a hit. In the classical world, one can determine which slit the electron has gone through; in the quantum world, one cannot know for certain.

naled a hit. If the electron behaved like a classical object, a bullet, for example, we could easily determine which hole it has passed through. Indeed, that is how police ballistics experts earn their living. But if the electron obeys quantum laws, as in fact it does, there is no way, *even in principle,* to determine which hole it actually passed through. Quantum physics determines only the probabilities that the electron has passed through one or the other of the two holes.

I should elaborate on this important point. With our everyday classical intuition, we can easily imagine finding out which route our motorist actually took by stationing several spies along the roadside of all possible routes. So, why can't we simply install a device by each of the two holes, a device which would click when an electron goes by? The answer is that, in the realm of the quantum, the very act of spying on the electron disturbs the electron so much that, in our example, the electron would no longer reach detector number 5 but would end up somewhere else entirely.

CERTAINTY ABOUT UNCERTAINTY

This drastic and unavoidable disturbance on the electron exemplifies the so-called uncertainty principle, which is sometimes explained by saying that when we observe a system we must disturb it. This statement, in itself, is a piece of triviality understood by a child taking a toy apart. It would imply that there is an uncertainty principle in classical physics when in fact there is not. The uncertainty principle is in fact considerably subtler.

In classical physics, nothing prevents us from making the disturbance on the system as small as we like. To stress this point by exaggeration, let me return to our example of the motorist. When operatives are sent out to ascertain which route the motorist actually takes, they may simply lay mines on the various highways and find out later which one actually exploded. But nothing prevents them from trying the subtler method of arranging for photons to bounce off the motorist's car into their eyes.

When we study the microscopic domain, the impact of a photon on an electron is necessarily considerable. This follows because each particle now is described as a probability wave. The reader might recall that to explain energy quantization, Prince de Broglie had to hypothesize that the wavelength of the probability wave of an electron in an atom is related to the momentum, and, hence, to the energy of the electron. The shorter the wavelength, the more momentum the electron has. Herein lies the crux of the uncertainty principle.

To find out where an electron is, we have to look at it. In other words, we have to arrange to have another particle, a photon, for example, bounce off the electron into some kind of detector. The accuracy with which we can determine the position of the electron is limited by the wavelength of the probability wave associated with the photon. The more accurately we want to determine the position of the electron, the shorter we have to make the wavelength of the photon. Then, according to de Broglie, the photon would, necessarily, have more momentum. Thus, an accurate measurement of the electron's position implies, inevitably, that we would have a very poor knowledge of its momentum after the measurement. The uncertainty principle does not say that we cannot measure either position or momentum as accurately as we like; it says that we cannot accurately measure both. The more accurately we measure the position of a particle, the less accurately we can measure its momentum, and vice versa. Remarkably enough, quantum physics is able to give a precise mathematical accounting of its own inherent uncertainties.

DICE UNLIKE ANY DICE

Welcome to the strange world of the quantum, where one cannot determine how a particle gets from here to there. Physicists are reduced to bookies, posting odds on the various possibilities.

That dynamics is probabilistic, rather than deterministic, is by no means the only strange thing about the quantum world. Indeed, as a physicist, I find it strange that popular discussions of the quantum world often stop at this point. The Good Lord not only plays dice, but He plays with very strange dice. Let me explain.

When a die is thrown, the probability of getting a 1 is $\frac{1}{6}$. The probability of getting a 2 is, of course, also $\frac{1}{6}$. Now, consider the following question: What is the probability of getting *either* a 1 *or* a 2 in one throw? The answer is obvious to gamblers and non-gamblers alike: The probability is $\frac{1}{6} + \frac{1}{6} = \frac{1}{3}$. In everyday life, to obtain the probability of either A or B occurring, we simply add together the probability of A occurring and the probability of B occurring.

The quantum die is astonishingly different. Suppose we are told that on the quantum die the probability of throwing a 1 is $\frac{1}{6}$, and the probability of throwing a 2 also $\frac{1}{6}$. In contrast to what our experience with ordinary dice might suggest, we *cannot*, in fact, conclude that the probability of getting either a 1 or a 2 in one throw is $\frac{1}{3}$! It turns out that the probability of throwing either a 1 or a 2 can range between $\frac{1}{3}$ and 0!

To say that the probability of an event occurring is 0 is to say that the event never occurs. Our intuition is outraged. The probability of throwing a 1 is ⅙, and the probability of throwing a 2 is also ⅙, yet the probability of throwing either a 1 or a 2 can be 0. How can that be? It does not make any sense! It doesn't, if by sense we mean the common sense we build up from living in the macroscopic world. The quantum world is truly strange.

In the quantum world, to obtain the probability of either A or B occurring, one does not add the probability of A occurring to the probability of B. The rule is more complicated than that. While the reader certainly does not have to master the rule of the quantum, I will state it to give you a flavor of what is involved. The law of the quantum actually determines a quantity known as the "probability amplitude" that a given event will occur. To obtain the probability that the event will occur, one squares the probability amplitude. The rule is that one adds the probability amplitude of A occurring and the probability amplitude of B occurring, to obtain the probability amplitude of either A or B occurring. One then squares the probability amplitude of either A or B occurring, to obtain the probability of either A or B occurring. One adds probability amplitudes, not probabilities.

ENTER THE ACTION

The fundamental law of quantum physics specifies what the probability amplitude is for each possible chain of events. Let us once again consider the basic problem of describing a particle moving from a point, call it "here," to a point x, in a specified period of time. (In physics, once this basic problem is mastered, then one can go on to formulate the more general problem involving the movement of many particles and fields.)

Recall that, in classical physics, the path of least action is followed. In quantum physics, we can only specify the probability amplitude of a given path being followed. Remarkably, the theoretical construct of "action" continues to be of central importance in quantum physics. The fundamental law of quantum physics states that the probability amplitude of a given path being followed is determined by the action corresponding to the path.

To find the probability amplitude, and hence, the probability of the particle getting to x by any path, we are instructed by our previous discussion to add up the probability amplitudes, one from each possible path. Thus, if we know in general that the particle is "here," we can determine that the probability amplitude for the particle will arrive at x at a later time.

While we can no longer predict, as we could in classical physics, exactly where the particle will go, we *can* predict the probability that it will get to any given point. It is not strictly correct, therefore, to say that quantum physics is nondeterministic. Rather, absolute determinism is replaced by a sort of gambler's determinism. A gambler cannot predict what number will come up with the next throw of the die, but he can predict that, after a great number of throws, the number 1 will come up about ⅙ of the time. Similarly, a quantum physicist can predict precisely the average value of many measurements.

One can easily understand why classical physicists such as Einstein were deeply disturbed. To predict the motion of a particle, we have to "read" all possible future histories of the particle, then "add" them up. The truth turns out to be stranger than our wildest imaginings.

SUM OVER HISTORY

The formulation of quantum physics just discussed is known, variously, as the "sum-over-history" or "path-integral" formulation. (Most textbooks and most popular expositions follow the wave-mechanical formulation invented by Erwin Schrödinger or the matrix formulation invented by Werner Heisenberg in 1925–1926.) The path-integral formulation was initiated by Paul Dirac, and developed around 1950 by Richard Feynman. One of its advantages is that the action is involved directly, thus making clear the connection between classical and quantum physics.

The path-integral formulation is ideally suited to discuss symmetry. If classical physics possesses a symmetry, then the action is invariant under certain symmetry transformations, as we learned in Chapter 7. It follows that since the same action controls quantum physics, quantum physics possesses the same symmetries as classical physics. For this and other reasons, over the last ten to fifteen years the path-integral formulation has largely supplanted the older wave-mechanical and matrix formulations in discussions of fundamental physics.

In the path-integral formulation, the essence of quantum physics may be summarized with two fundamental rules: (1). The classical action determines the probability amplitude for a specific chain of events to occur, and (2) the probability that either one or the other chain of events occurs is determined by the probability amplitudes corresponding to the two chains of events.

Finding these rules represents a stunning achievement by the founders of quantum physics. The mental processes involved can only be described as quantum leaps of genius.

The law of the quantum is not so much a theory in itself, but a prescription to obtain a theory relevant in the realm of the quantum. One obtains quantum mechanics by applying the prescription to Newton's theory of mechanics, quantum electrodynamics by applying it to Maxwell's theory of electromagnetism, quantum gravity by applying it to Einstein's theory of gravity. But the action of quantum electrodynamics is still Maxwell's action, with all its symmetries.

It is sometimes said that quantum theory is nothing but a recipe or prescription for arriving at predictions that are to be compared with experimental observations. This statement misses the point. Actually, one can argue quite validly that physics itself is a collection of recipes for obtaining predictions that accord with experiments. Newton's laws form a recipe in the same sense that the quantum laws, as summarized above, amount to a recipe. It would be circular reasoning to say that Newton's laws are better "understood" because they are more in accord with our everyday intuition. What is true is that Newton's laws can now be understood as an approximation of the quantum laws under certain circumstances. Perhaps one day we will discover that the quantum laws are themselves approximations of a more fundamental set of laws. What physicists hope for is that our present recipe can one day be derived from another recipe, more elegant and concise in form, and more universal in applicability. Again, ultimately there is no "why," only "how," in our dialogue with Nature. Theoretical physicists try to know Her thoughts, but as far as I can see, they will never know why She thought those thoughts.

HERE AND THERE AT THE SAME TIME

The profound difference between classical and quantum physics is underscored by how we describe the "state" of a system at a given point in time. The notion of the state of a system is a natural one. For instance, the President of the United States is required to give a "State of the Union" address every year. For simplicity, let us consider a single particle, the electron.

In classical physics, the state of the electron is specified by its position at the given instant. For instance, if the electron is in Paris, we would say that the electron is in the state |Paris>. Or, if the electron is in Rome, we would say that the electron is in the state |Rome>. (Traditionally, physicists denote a state by the symbol |name>, where "name" specifies which state we are talking about.)

When we pass into quantum physics, we can no longer specify the position of the electron. Instead, the state of the electron is specified by a probability amplitude that tells us the probability of finding the electron at any location in space, at the given time. For instance, the electron may be in the state |"Paris">, specified by saying, for example, that the probability amplitude is $\frac{1}{2}$ that the electron is in Paris, $\frac{1}{10}$ that it is in Rome, and so on. Since the electron is most likely to be found in Paris, we continue to name the state Paris, but we use quotation marks to remind ourselves that we can only post odds on the location of the electron. Similarly, another possible state for the electron to be in may be called |"Rome">, specified by saying, for example, that the probability amplitude is $\frac{1}{2}$ that the electron is in Rome, $\frac{1}{10}$ that it is in Paris, and so on. The job of the quantum physicist is to classify all possible states, and to determine, as time flows on, how the electron can leap from one state to another.

We are interested, however, in another point here. In the strange realm of the quantum, we can add states! For example, we can consider the state |"Paris"> + |"Rome">, a state specified by saying that the probability amplitude is $\frac{1}{2} + \frac{1}{10}$ that the electron is in Paris, $\frac{1}{10} + \frac{1}{2}$ that it is in Rome, and so on. In fact, we can add two states in any proportion we like. Thus the state $a|$"Paris"$> + b|$"Rome"$>$, with a and b denoting two numbers of our choice, is specified by saying that the probability amplitude is $a \times \frac{1}{2} + b \times \frac{1}{10}$ that the electron is in Paris, $a \times \frac{1}{10} + b \times \frac{1}{2}$ that it's in Rome, and so on.

That states can be added together is another truly bizarre feature of the quantum world. In classical physics, it makes absolutely no sense to add two states together. What could the state |Paris> + |Rome> mean? Classically, the electron cannot be in Paris and Rome at the same time.

INTO THE LEADING ROLE

Finally, the stage is set for explaining how symmetry found stardom in the quantum world. It is the possibility of adding states together that makes symmetry considerations more powerful in quantum physics than in classical physics. To be specific, let us discuss rotational symmetry.

Consider a planet orbiting around a star. What does rotational symmetry tell us? Not, as we saw before, that the orbit must be a circle, but, rather, that if we rotate the orbit through any angle we choose, the rotated orbit is a possible orbit. See Figure 2.2, page 12. This conclusion is rather obvious, and not particularly interesting.

In contrast, consider an electron orbiting around an atomic nucleus. We expect rotational invariance, described by the group $SO(3)$, to hold. We are now in the quantum realm and forbidden to speak of precise orbits. Instead, we can only speak of the state of the electron. Put a quantum theorist to work and have her classify the possible states of the electron around the nucleus. Suppose the electron is in the state |1>.

Let us rotate the atom through some angle of our choice, and denote its new state by |R1>. Rotational symmetry, by definition, tells us that |R1> is a possible state, and, furthermore, that it must be a state with the same energy as |1>. To see that this follows, we may consider rotating the observer rather than the atom. Better yet, let us compare the perceptions of two observers whose viewpoints are related by the rotation, as described in Chapter 2. One observer sees the state |1>, the other the state |R1>. To say that physics is rotationally invariant is to say it does not prefer one observer over the other. Thus, the state |R1> and the state |1> must have the same energy.

At this point, two logical possibilities present themselves to our discerning minds: Either the state |R1> is exactly the state |1>, or it is not.

Suppose $|R1>$ is equal to $|1>$. This says that when we rotate the atom, the electron remains in the same state. The probability of locating the electron is unchanged by the rotation; in other words, the probability distribution of the electron in the state $|1>$ is spherically symmetric. Observers whose viewpoints are rotated from each other see the same state $|1>$.

The second case is more interesting. $|R1>$ is not equal to $|1>$. In general, $|R1>$ may be the sum of $|1>$ and some other states. For the sake of definiteness, let us say that four other states, labeled $|2>$, $|3>$, $|4>$, and $|5>$, are involved. In other words, $|R1>$ may be equal to the linear combination $a|1> + b|2> + c|3> + d|4> + e|5>$. (The numbers a, b, c, d, e depend on which rotation we are talking about, of course.)

Now, suppose we rotate the state $|2>$. Applying the same reasoning as above, we may expect the rotated state $|R2>$ to be equal to the linear combination $f|1> + g|2> + h|3> + i|4> + j|5>$, with the numbers f, g, h, i, j depending on the rotation in question. We can go on and rotate the states $|3>$, $|4>$, and $|5>$, and each one of the rotated states will be equal to a linear combination of the five states $|1>$, $|2>$, $|3>$, $|4>$, $|5>$.

A BELL RINGS

Now a bell rings in the back of our minds. This discussion seems rather familiar. Indeed, the situation here is precisely the same as the one we encountered while discussing group representations. Here, under rotations, the quantum states $|1>$, $|2>$, $|3>$, $|4>$, and $|5>$ are transformed into linear combinations of themselves. They furnish a five-dimensional representation of the rotation group $SO(3)$.

In the preceding chapter, we spoke of abstract entities, or arrows, or whatever, transforming into linear combinations of each other. Remarkably, the abstract mathematical discussion of the nineteenth century is realized physically in the transformation of quantum states. The intrinsic mathematical structure does not depend on whether we are talking of abstract "entities" or quantum states, apples or kittens.

For the sake of definiteness in our discussion, I supposed that $|1>$ belongs to a five-dimensional representation. In general, the quantum state $|1>$ could belong to a representation of whatever dimension allowed by the group. For instance, it might belong together with eight other states, to a nine-dimensional representation. Which representation a given state actually belongs to depends on detailed physics.

GROUP THEORY IN QUANTUM PHYSICS

What does this discussion of symmetry and group theory in quantum physics actually imply for experimental observations?

We learned that the quantum states of an electron in an atom belong to representations of the rotation group. Rotational symmetry tells us that the states belonging to the same representation all have the same energy. As I have already indicated, this follows because these states can be rotated into each other. Thus, in our example, we could have chosen a rotation under which $|1>$ is rotated into $|2>$—in other words, a rotation such that $|R1>$ is equal to $|2>$. Indeed, experimenters have observed different quantum states having exactly the same energy.

Recall that the dimensions of the allowed representations are fixed by group theory. For example, the rotation group does not have a four-dimensional representation. If experimenters, therefore, observe a set of four quantum states of the same energy in our atom, they would know from group theory that they must be able to find additional states of the same energy.

Experimentally, the energies of the electron states in an atom are inferred from the energy of the radiation emitted when the electron leaps to a state of lower energy. Suppose the electron leaps from a state belonging to a five-dimensional representation to one belonging to a representation that is seven-dimensional. There are in total $5 \times 7 = 35$ (ordinary multiplication!) different leaps possible. Without group theory, atomic physics would become an extremely tedious subject, in which each of these 35 possible leaps, in turn, would have to be studied. But rotational symmetry and group theory can tell us immediately what the relative probability of each of these 35 leaps is without a tedious calculation. (The intensity of the radiation emitted in a given leap, experimentally, is directly proportional to the likelihood of that leap.) As I have emphasized in Chapter 2, basically rotational symmetry simply requires two observers, one with his head tilted relative to the other, to perceive the same structure of physical reality. This apparently innocuous requirement is powerful enough to fix the relative probability of each of the 35 possible leaps.

Incidentally, the probability for some leaps may be forced by group theory to be zero. Rotational symmetry, in other words, forbids the electron to take that particular leap. Physicists call this a selection rule. In general, a certain number of quantum transitions appears to be possible, a priori. But an underlying symmetry will allow only certain transitions to proceed. The others are taboo.

Selection rules are in fact manifestations of the connection between symmetry and conservation. According to Emmy Noether, the presence of a symmetry implies a conservation law. Just as processes that do not conserve energy are forbidden, certain quantum leaps are forbidden because they violate a relevant conservation law.

Historically, physicists studying atoms were confronted with a confusing morass of experimental data. Many states are of the same energy. Of the numerous possible leaps between states of one energy and states of another energy, some occur more often than others. The eminent Hungarian-American physicist Eugene Wigner realized finally that, with rotational symmetry and group theory, order could be wrought from the chaos.

THE TRIUMPH OF SYMMETRY IN THE QUANTUM REALM

Let us pause and take stock of what we have learned. In classical physics and in quantum physics, symmetry restricts the possible forms of the basic laws. But in quantum physics, symmetry goes further. While the notion of adding two different orbits makes no sense in classical physics, we are entitled to add quantum states, thanks to the probabilistic interpretation of quantum physics. Under a symmetry transformation, the transformed state may be a linear combination of quantum states. In a reversal of the abstraction process, the ruminations of nineteenth-century group theorists are realized in the quantum world. (It is as if a civilization had figured out the theory of multiplying numbers, only to realize later that the rules could actually be applied to situations involving baskets and apples.) If the symmetry is worth its very name, the states belonging to the same representation must have the same energy. Symmetry then regulates the quantum leaps between quantum states. Thus, in quantum physics, symmetry not only tells us about the underlying laws, it also tells us about the actual physical states.

IV

TO KNOW HIS THOUGHTS

11

The Eightfold Path in the Forest of the Night

TWINS IN THE SUBNUCLEAR FOREST

When Alice ran into Tweedledum and Tweedledee, she was pleasantly intrigued. In 1932, a compatriot of Alice's, James Chadwick, while wandering through the newly opened-up nuclear forest, encountered a Tweedledum and Tweedledee of his own. As a result of this remarkable encounter, Chadwick later was knighted.

Chadwick, whom we met in Chapter 3 as a hapless prisoner of war, discovered a hitherto unknown particle, the neutron, which behaves exactly like a proton as far as the strong nuclear force is concerned. Since Chadwick's discovery, physicists have discovered that the subnuclear population contains not only identical twins, but also identical triplets, even identical octets. Like Alice, physicists have been puzzled and intrigued. What is Nature trying to tell us?

By 1930, physicists had begun to study the atomic nucleus. The exploration was made possible by an extraordinary kindness on Nature's part: She provided just the necessary tool in the form of naturally radioactive substances. Radioactivity had been discovered accidentally in 1896 by the French physicist Antoine Henri Becquerel.

Radioactive substances, it was soon understood, contain unstable nuclei that seek to rearrange themselves and, in the process, shoot out particles of various kinds. I have already mentioned that the process of looking at an object consists of bombarding the object with photons and catching the scattered photons with those marvelous optical detectors we carry in our heads. Particle accelerators are simply gargantuan devices built to extend the basic process of seeing. To look at the inner structure of matter, we have to bombard matter with particles energetic enough to penetrate the outer layers of matter. Radioactive substances provided a natural source of energetic particles at a time when physicists had not yet had the idea of building accelerators. Taking advantage of these natural "accelerators," physicists began to expose various materials to known radioactive sources.

In 1930 the German physicists W. Bothe and H. Becker discovered that certain materials, when exposed to radioactive sources, emitted a mysterious radiation. At that time, physicists believed that the world was made of electrons, protons, photons, and gravitons. It was understood that an atom of matter consisted of electrons orbiting about a nucleus. The nucleus was thought to be made of protons and, possibly, also electrons. Puzzled by the German report, Chadwick performed a series of experiments to show that this mysterious radiation consisted of a hitherto unknown particle. The particle was electrically neutral and so became known as the neutron.

Observing the collision of a tennis ball and a golf ball, we can easily determine the relative mass of the two by invoking the conservation of energy and momentum. By careful observation of the neutron colliding with various atomic nuclei, Chadwick managed to measure the mass of the neutron in the same way. To his surprise, the mass of the neutron came out to be almost exactly the same as that of the proton. The neutron plays Tweedledee to the proton's Tweedledum.

NOT MERELY AN IDLE CAMP FOLLOWER

Further experiments quickly established that an atomic nucleus is composed of a certain number of protons and neutrons. What had happened in the experiments of Bothe and Becker was that the energetic emission from the radioactive source had knocked out some of the neutrons.

Chemical properties of an atom are determined by the number of electrons orbiting outside the nucleus. The number of electrons equals the number of protons so that the atom, as a whole, is electrically neutral. Thus, the neutron plays no role in the chemical properties of the atom. For example, carbon atoms always contain six protons. The fact that it has six protons, and not five or seven, is what gives a carbon atom its "carbon-ness," which includes its unique propensity to bond and hence its essential role in biology. But carbon atoms have been observed to have anywhere from four to nine neutrons. What role has been assigned to the neutron in the drama of physics? Is the neutron merely an idle camp follower around the mighty proton? Hardly. It turns out that, without neutrons, atomic nuclei would not be stable.

A BALANCING ACT

The stability of atomic nuclei, and by extension, the stability of the entire world, hinges on a high-wire balancing act of Nature. Since protons are electrically charged, they repel each other. This electric repulsion between the protons in a nucleus threatens to tear the nucleus apart. Thus, the very existence of the nucleus compelled physicists to conclude that the protons and neutrons inside the nucleus are held together by a strong mutual attraction. Physicists refer generically to protons and neutrons as nucleons. The new interaction responsible for the attraction between nucleons is called the strong interaction, since it turns out to be about a hundred times stronger than the electromagnetic interaction.

One might think, therefore, that the electric force, being so much weaker, would be totally overwhelmed in the nucleus. But Nature has thrown in an interesting twist. The electromagnetic interaction, though weaker, has a longer reach. Recall that the electric force between two charges decreases as the square of the distance separating them. The strong interaction between two nucleons, in contrast, decreases so much more precipitously that two nucleons attract only when they are practically next to each other. The strong interaction is said to be short-ranged, the electromagnetic interaction long-ranged. At a crowded cocktail party, one can chat via the short-range acoustic interaction only with those persons one is standing next to, but one can wink at an attractive stranger clear across the room via the long-range optical interaction.

The nucleus may be thought of as a bag of nucleons. The nucleons are strongly attracted to each other, but each can only tug at those nucleons next to it. The electric repulsion, while much more feeble, can reach from one proton to another clear across the nucleus. The atomic nucleus provides the arena for an interestingly matched prizefight. One boxer has a stronger punch but a shorter

Figure 11.1. The atomic nucleus provides the arena for an evenly matched prizefight.

reach, while his opponent has a longer reach but a weaker punch. Electric repulsion, evidently, tends to win out in a large nucleus. For example, the uranium nucleus, with its 92 protons and its 140 (or so) neutrons, is prone to fission at the drop of a hat. The electric repulsion tears the nucleus apart and in the process liberates a certain amount of energy that we humans have tried to put to a variety of uses, some more sensible than others. At the other end of the spectrum, two small nuclei can be persuaded to fuse together. According to some people, the energy liberated in this fusion process will be essential to the future of the human race.

Fortunately for us, the strengths of the strong and electric forces are such that a wide variety of perfectly stable nuclei exist in which neither force is able to achieve a knockout. The neutron plays an essential role in maintaining the stalemate. In a stable nucleus, the neutrons, being electrically neutral, could help out the strong force without adding to the electric repulsion. The helium nucleus, for instance, has two protons and two neutrons. If the neutrons were not there, the helium nucleus would fall apart. As explained in Chapter 2, we are able to bathe in the steady warm glow of the sun precisely because of this sort of balancing act performed by Nature.

It is remarkable that the visible structure of the physical world depends on the essential presence of all the fundamental interactions. Were the strong interaction absent, nuclei would not exist, and the only possible atoms would be those of hydrogen, formed of a proton and an electron. The universe would consist of just a gas of hydrogen and some neutrons floating about freely. Were the electromagnetic interaction absent, atoms would not exist, and the universe would contain lumps of nucleons with the electrons floating about freely. When two lumps met, they would stick to each other to form a bigger lump. All the matter in the universe might have ended up in one big lump.

THE WEAK INTERACTION

During the 1930s, it became increasingly clear that yet another hitherto unknown interaction, the weak interaction, was responsible for the radioactivity of certain kinds of nuclei. The discovery of the weak interaction rounds out the list of fundamental interactions known to physics. We have already encountered the weak interaction in connection with parity violation, and in a later chapter, the weak interaction will be discussed in more detail. Here, I will only mention that the range of the weak interaction is a thousand times shorter than that of the strong interaction. It is because of their short ranges that the strong and weak interactions do not manifest themselves in macroscopic phenomena, as opposed to the electromagnetic and gravitational interactions, which are both long-ranged.

NATURE REVEALS A SYMMETRY

I have now restored to the neutron its self-respect by explaining how it is essential to the healthy functioning of the universe. That still leaves the mystery of why the neutron is so close in mass to the proton. Nothing in the preceding discussion requires the proton and the neutron to have the same mass. The mass of the proton and of the neutron has been measured to be about 938.2

MeV and 939.5 MeV, respectively. The difference is only about ⅒ of a percent! (Incidentally, 1 MeV, that is, 1 million electron volts, is the energy acquired by an electron accelerated through a voltage drop of 1 million volts. Many physicists customarily measure mass in energy units since Einstein abolished the distinction between mass and energy.)

Further studies revealed yet another surprising fact: The strong force between two protons, between a proton and a neutron, and between two neutrons was measured to be approximately the same. The neutron behaves just like the proton, except for the almost negligible fact that one is charged, the other is not —negligible, because the electromagnetic force is so much weaker on an individual nucleon.

Here we have Tweedledum and Tweedledee: They talk and act like each other, they weigh the same to within ⅒ of a percent, but one of them has a mustache and the other not.

In 1932, Werner Heisenberg, certainly not known as a conservative physicist, boldly proposed that Nature is dropping a very loud hint that the neutron puzzle can only be understood in terms of a fundamental symmetry in Nature's design. Heisenberg began by imagining what would happen if he could switch off the electromagnetic, the weak, and the gravitational interactions. I have already mentioned in Chapter 2 the very useful trick of switching off, or neglecting, the more feeble interactions when studying a given interaction. Heisenberg guessed that the neutron and the proton would become exactly equal in mass. This guess, that electromagnetism is responsible for the tiny proton-neutron mass difference, is not unreasonable. Since the electromagnetic force is about a hundred times weaker than the strong force, one might naively expect its effect to be about 1 percent or less.

In the preceding chapter, I explained that various atomic states may be rotated into each other. Rotational symmetry guarantees that these states have exactly the same energy. Recall, the argument is simply that two observers whose viewpoints are rotated from each other must deduce the same structure of physical reality. This, after all, is the trivial *and* profound essence of the symmetry argument. Inspired by rotational symmetry, Heisenberg postulated that the proton may be "rotated" into the neutron, and that the strong interaction is invariant under this rotation.

Heisenberg's logic is the reverse of that used in our discussion of rotational symmetry and atomic states. Since the days of

Newton, our intuition practically requires rotational symmetry. With quantum physics, rotational symmetry implies that various atomic states must have the same energy. Heisenberg, on the other hand, started with the surprising discovery of Chadwick's that the proton and the neutron have almost the same mass (and therefore the same rest energy, according to Einstein) and deduced the presence of a hidden symmetry of Nature's design.

In Chapter 6, we asked how physicists wanting to play Einstein's game could ever get to square one. Here is an example in which an experimental fact sings of a symmetry to those who can hear.

Heisenberg's symmetry is called "isospin," for various historical reasons that do not concern us here. The corresponding group is called by mathematicians $SU(2)$. The number "2" reminds us that the group is defined by transforming two objects into each other.

I must interject at this point that in recounting the introduction of isospin I have sacrificed historical accuracy in order not to interrupt the narrative flow. As is more often the case than not in the history of physics, the development of isospin was full of misconceptions and confusions. Many physicists contributed to the clarification of isospin as a symmetry. I find it convenient, however, to attribute isospin solely to Heisenberg. In this, I am indulging in the same kind of bowdlerization of history of which even physics textbooks are usually guilty. A brief sketch of the history of isospin is presented in a note to this chapter.

VIEW INTO AN INTERNAL WORLD

Heisenberg's isospin rotation is not a rotation like those in the actual space we live in; hence the quotation marks in a preceding paragraph. Rather, Heisenberg envisions a rotation in an abstract internal space; the terms "rotation" and "space" are both used figuratively.

Isospin represents a stunning landmark in the development of symmetry as a primary concept in physics. Previously, when physicists thought of symmetry, they thought of the symmetry of spacetime. Parity, rotation, even Lorentz invariance and general covariance, are all rooted, to a greater or lesser degree, in our direct perception of an actual spacetime. Now, in one sweeping motion, Heisenberg opened up for us an abstract inner space in which symmetry operations can act also.

The Old Guard must have found Heisenberg's proposal hard to take. The symmetries of spacetime always had been thought unquestionably exact. But here comes Heisenberg, proposing a symmetry that is manifestly *not* exact. The symmetries of spacetime are universal: They apply to all interactions. Isospin symmetry applies only to the strong interaction: The proton and the neutron have different electromagnetic properties.

With the passage of time, Heisenberg's notion of an internal symmetry no longer appears so revolutionary. To later generations of physicists, internal symmetry seems as natural and real as spacetime symmetry.

I have emphasized that symmetry principles tell us that physical reality, though perceived to be superficially different by different observers, is in fact one and the same physical reality at the structural level. In the present instance, one observer sees a proton, but another observer, whose viewpoint is isospin rotated from the first, may insist that he sees a neutron. They are both right, in exactly the same way that what is "up" to one observer may be "down" to another. The observed fact that the strong force between two protons is the same as that between two neutrons follows immediately, since what looks like two protons to one observer looks like two neutrons to another.

THE FULL FORCE OF GROUP THEORY

Once the observed Tweedledum-Tweedledee situation is formulated as a symmetry, then the full force of group theory may be brought to bear on the physics. One can either work out the representations of $SU(2)$ or consult a mathematics book. The general considerations in Chapter 7 imply that any particle which interacts strongly, from atomic nuclei to various subnuclear particles, must belong to a representation of $SU(2)$. The particles belonging to the same representation are said to be members of a multiplet; more specifically, a doublet, triplet, quartet, and so on. All members of a multiplet must have the same energy or mass. This is indeed observed.

According to Emmy Noether, a conserved quantity must be associated with isospin symmetry; it is called simply isospin. Particles that interact strongly carry isospin in much the same way that particles that interact electromagnetically carry an electric charge. Strong interaction processes that do not conserve isospin are forbidden. Furthermore, the relative probabilities of various allowed processes are determined by group theory. The situation is entirely analogous to that encountered when we discussed rotational symmetry in the preceding chapter, and necessarily so, since the controlling mathematics exists independent of physics.

STRONG INTERACTION IS TOO STRONG

Once the physical idea of isospin symmetry is understood, the detailed applications that follow are not particularly relevant to our story and are best left to professional nuclear physicists. The important point is that the experimentally verifiable predictions in the preceding discussion are strictly consequences of isospin symmetry alone. I never mentioned what the theory of the strong interaction might be. It does not matter!

If one were to try to construct a theory of the strong interaction, it is true that isospin would severely restrict the possible form of the theory. But even if one has a theory, it is not of much use, since strong interaction is, by definition, strong. Let me explain.

Physics students often get the impression from textbooks that physics is concerned with exact solutions. To illustrate various physical principles, textbook authors naturally tend to treat those simple and idealized cases for which exact solutions are possible. In actual practice, physicists have to resort to a method known as perturbation. For example, to work out the motion of the earth around the sun, a physicist would start by ignoring the other planets, then he would calculate the effects of the other planets on the earth's orbit approximately. This procedure works well because the effects of the other planets are small.

The basic idea is similar in quantum physics. When we scatter two electrons, the probability that the two electrons will interact is only about $1/137$. This empirical number, $1/137$, measures the strength of the electromagnetic interaction, and it is known as the electromagnetic coupling constant. Suppose the two electrons do interact. As they are moving apart, there is a quantum probability that they might interact again. The probability that the electrons would interact twice is ($\frac{1}{137}$) × ($\frac{1}{137}$), about one chance in ten thousand. We could thus either neglect the effect of double interaction or treat it as a small correction. Fortunately for physicists, three of the four fundamental interactions have weak couplings and the perturbation method can be used.

Nature is kind—but not kind enough. In the strong interaction, the coupling constant is essentially 1. Therefore, when we scatter two nucleons, double interaction, triple interaction, and so on ad infinitum are all just as likely to occur as a single interaction! Here perturbation fails utterly. The annals of physics are full of futile attempts to calculate the force between two nucleons from first principles. Nuclear physicists eventually gave up and adopted a quasi-phenomenological approach, taking the experimentally measured force between two nucleons as a given, then trying to calculate the properties of nuclei.

Eminent football players are often known by their nicknames: Harvey "Too Mean" Martin and Ed "Too Tall" Jones, for example. To theoretical physicists, the strong interaction is "too strong" and "too mean."

CONTAINED IGNORANCE

The power and glory of symmetry allow us to bypass completely the construction of strong interaction theories of dubious utility. We are able to contain and isolate our ignorance.

Historically, this containment of our ignorance was of considerable importance. Most particles participate in more than one interaction. (For example, the proton participates in all four fundamental interactions.) In studying the weak interaction, physicists encounter many processes involving particles that also interact strongly. Fortunately, by using symmetries, physicists concentrating on the weak interaction were able to contain the strong interaction monster. The structure of the weak interaction, as a result, was completely elucidated by the early 1970s; physicists did not have to wait for a complete theory of strong interaction.

The situation with the strong interaction may be seen in Figure 11.2.

Of course, a complete theory would tell us more than would symmetry considerations alone. Symmetry tells us that states in

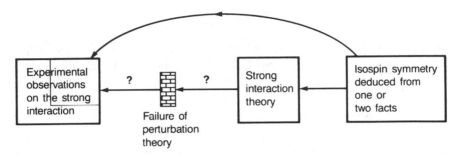

AN END RUN STRATEGY

Figure 11.2.   The schema of strong interaction research from the 1930s to the early 1970s. Because the strong interaction is "too strong," physicists were unable to extract experimental predictions from a theory of strong interaction, even if they were able to construct one. Symmetry considerations allow them to bypass the roadblock, represented here as a brick wall, set by the failure of the perturbation method.

the same multiplet must have the same energy, but it cannot tell us what that energy is. Strictly speaking, therefore, the path in Figure 11.2 coming out of symmetry and bypassing the theory box could only lead us to some of the experimental observations. We have indicated this as a smaller box set inside the experimental observations box.

THE MARRIAGE BROKER

While the strength of the strong interaction inhibits the construction of a meaningful theory, it does not prevent us from understanding the nature of the interaction. In the early 1930s, the Japanese physicist Hideki Yukawa wondered why the strong interaction was short-ranged, in contrast to the two better-known interactions.

As we noted in Chapter 4, the notion of field replaced action at a distance. We picture the electric force between two charges as resulting from each charge "talking to" the electric field sent out by the other charge. In quantum physics, the energy in the field is concentrated in packets, photons in the case of the electromagnetic field, gravitons in the case of the gravitational. Thus, according to modern physics, when two electrons are present, one of them would emit a photon with a certain probability amplitude, and the other would absorb the photon. The process repeats itself rapidly. This constant exchange of photons between the two electrons produces the observed electric force. Similarly, the constant exchange of gravitons between our bodies and the earth keeps us earthbound. Like the marriage brokers of old and the shuttle diplomats of new, photons travel tirelessly between two parties, telling each the other's intentions.

Since the early days of physics, the notion of force has been among the most basic and the most mysterious. It was thus with considerable satisfaction that physicists finally understood the origin of force as being due to the quantum exchange of a particle.

Given this understanding of the nature of force, in 1934 Yukawa decided that a "marriage broker" must be provided for the strong interaction as well. Boldly, Yukawa hypothesized a new particle, which became known later as the pi meson, or the pion for short. Strikingly enough, he was able to predict the properties of the pion, a feat for which he was awarded the Nobel Prize.

Consider two nucleons sitting more or less at rest inside an atomic nucleus. One of them emits a pion with a certain probability amplitude, and the other absorbs the pion. Like a marriage broker, the pion shuttles back and forth between the two nucleons. Focus on one of the nucleons emitting a pion. Wait! Something seems to be wrong. The proposed emission process would violate energy conservation! According to Einstein, even a particle sitting at rest carries a certain amount of energy, equal to its mass times c^2. How can a nucleon, sitting inside a nucleus and having an energy equal to its mass times c^2, emit a pion, which, even if it is barely moving, must have at least as much energy as *its* mass times c^2, and remain a nucleon?

Figure 11.3. The pion as a marriage broker of old, a corpulent lady whose inclination to travel is limited by her weight.

Remarkably, Yukawa managed to turn this apparent difficulty to his advantage. By an inspired use of the uncertainty principle, he was able to explain the short range of the strong interaction and to predict the mass of the pion. The key is quantum physics.

THE EMBEZZLER

I have explained that the uncertainty principle tells us that we cannot measure both the momentum and the position of a particle as accurately as we would like. Similarly, we cannot measure both the energy of a process and the time the process takes as accurately as we like. (Recall the relationship between physical concepts in Einstein's work: A particle's position in space is related to its position in time; its momentum is related to its energy. Therefore, it is plausible that the uncertainty principle governs momentum and position as a pair, and energy and time as another pair.) Yukawa realized that if we pinpoint the time at which the nucleon emits the pion, then we are uncertain about the energy involved and we can't tell if energy is conserved or not. The uncertainty principle allows energy not to be conserved, but only momentarily.

The situation reminds me of embezzling. A fundamental principle of embezzling says that the larger the amount of money stolen, the sooner the theft will be noticed. The pion is like an embezzler trying to make off, not with money, but with energy. Unlike embezzlers in real life, the pion is always caught by Nature and required to give back its energy. Nature, as represented by Emmy Noether, demands that the energy of an entire process, in this case the interaction between two nucleons, be conserved. Thus the pion must be absorbed quickly by a nucleon. As with embezzling, the more energy the pion tries to carry away, the quicker it is caught.

Nature is so vigilant that even if the pion is traveling at the speed of light, it cannot get very far between emission and absorption. Yukawa thus explained the short range of the nuclear force. If two nucleons are too far apart, they will not feel the presence of each other. Using the other analogy, I may say that perhaps the pion is like the marriage broker, a corpulent lady whose inclination to travel is limited by her weight. When the two parties live too far apart, the broker is reluctant to get involved.

The range of the nuclear force is clearly determined by the mass of the pion. The minimum energy the pion tries to steal away with is the rest energy associated with its mass. This minimum energy sets the maximum time allowed the pion to get from one nucleon to another. Knowing the range of the nuclear force, Yukawa predicted that the pion should have a mass about one tenth that of the nucleon.

Incidentally, we also understand now why the electromagnetic interaction is long-ranged, since the photon is known to be massless. While the graviton has never been detected, it is believed to be massless since the gravitational interaction is also long-ranged.

AN ARROGANT CONSERVATIVE

To predict that a fundamental particle with certain properties exists is a supreme act of the rational mind. Dirac and Pauli had done it, and a few years later, so did Yukawa. Their acts went defiantly against the social climate then prevalent in the physics community. Yukawa later wrote that to think outside of the known limits of matter was "to be arrogant, not to fear the wrath of the gods" and that there was "a strong inhibition of such thoughts that was almost unconscious."

There is an immense satisfaction in being arrogant with Nature and then in seeing Her oblige. Unfortunately for theoretical physicists, this is a pleasure much desired but seldom granted. These days, the inhibition of which Yukawa spoke no longer holds sway, and theorists are predicting new particles with wanton abandon. The situation, indeed, has so degenerated that some theorists of my generation are apt to invent new particles for no good reason other than to explore their consequences were they to exist.

Interestingly, Pauli's hypothesis of the neutrino and Yukawa's hypothesis of the pion, bold as they were in absolute terms, represent a relatively conservative stance. When faced with the mysteries of the nucleus, some eminent physicists of the time argued for a breakdown in the quantum laws. Classical mechanics, after all, had broken down at the atomic scale. It seemed reasonable to think that quantum physics would fail at the nuclear scale, tens of thousand times smaller than the atomic scale. In a phenomenon rarely if ever seen in political history, the revolutionaries who dethroned classical physics eagerly plotted to overthrow their own revolution.

THE POSTWAR BOOM

In the discussion thus far, we have been thinking of two nucleons sitting inside a nucleus. The lack of energy forces the pion to shuttle between the two. But if we collide two nucleons at high energies, the additional energy of motion carried by the nucleons may be sufficient to produce a pion while still conserving energy. The pion was discovered in precisely this way in the late 1940s.

The pion proved to be only the first of a horde of particles discovered after the war. During the postwar baby boom, experimental physicists were busily producing particles one after another. In 1947 came the first of the infamous "strange" particles, so called simply because physicists never expected them. (To my generation of physicists, there is nothing particularly strange about strange particles.)

Strange particles are produced when nucleons collide at high energies. Experimental studies showed that these previously unexpected particles are never produced singly, but always in pairs. For example, when a strange particle known as K^0 is produced, it is always accompanied by the particle Σ^+. Nuclear collisions never produce just K^0, or two K^0's, but always K^0 with Σ^+. A number of such empirical rules were accumulated.

Eventually, it became clear that all these empirical rules could be summarized as a conservation law. A new physical quantity, dubbed "strangeness," is supposed to be conserved by the strong interaction. Think of strangeness as analogous to electric charge. It is a physical attribute carried by some strongly interacting particles, but not by others, just as electric charge is carried by some particles, not others. The proton, the neutron, and the pion are supposed to carry zero strangeness. The newly discovered particles were assigned the various degrees of strangeness $+1$, -1, and so on. From that point on, the word "strange" took on a specific meaning for physicists.

Let us now see how strangeness conservation can explain

the actual observations. Let K^0 be assigned strangeness $+1$. Since the nucleons have zero strangeness, they cannot produce K^0, either singly or in a pair. Now, if Σ^+ carries strangeness -1, then we can account for production of K^0 and Σ^+ together.

The idea of strangeness conservation is a simple one, but it certainly was not clear, a priori, that the accounting scheme would work. For instance, if the scheme is to work, experimentalists had better not see Σ^+ produced, either singly or in pairs. In this way, the scheme was checked in numerous processes and was found to hold.

Once again invoking Noether's insight on the connection between symmetry and conservation, physicists immediately concluded that strangeness conservation signals a symmetry beyond isospin. I will return to this point shortly.

WHAT IS IN A NAME?

If I could remember the names of all these particles, I'd be a botanist.
—E. Fermi

Physicists had fun naming all the new particles. To divide particles into classes, they called particles that interact strongly "hadrons," whose Greek root means stout or thick. (Thus, a hadrosaur is a particularly gigantic dinosaur.) Nucleons, pions, and strange particles are all hadrons. The particles that do not interact strongly, such as the electron and the neutrino, are called "leptons," Greek for thin, delicate, small. (Thus, the lepton is the least valuable coin in Greece, and a person having a thin, narrow face is leptoprosopic.)

The hadrons were classified further. The pi meson, later shortened to pion, was so named because it is intermediate in mass between the nucleons and the electron. (The root "meso," meaning middle, is of course well known, as in mezzosoprano and Mesopotamia.) New particles with properties similar to the pion's are called mesons. In contrast, the nucleons and those new particles with properties similar to the nucleon's are called "baryons," Greek for heavy. (In music, we have the baritone.) The new terminology turned out to be inaccurate in some cases. We now know of some mesons that are more massive than some baryons. Even so, the name meson actually is an inspired choice, as it also suggests a mediator or go-between. In a fascinating, multilingual pun, Chinese physicists refer to the pion as 介 子. The Chinese character 介 means mediator, but it also happens to be reminiscent of the Greek letter for pi, π.

Since the nucleons are present in ordinary matter and therefore play a more important role, it is useful to distinguish them from the other baryons. Baryons were thus subdivided into nucleons and hyperons. The Σ^+ (sigma plus) mentioned above is a hyperon. There is also a hyperon named the xi, denoted by the Greek letter Ξ. Supposedly, the hyperons were named after the song "The Sweetheart of Sigma Chi," with a suitable corruption of the fraternity's name.

Speaking of names, I may also mention that neutrino was derived by adding the Italian diminutive "-ino" (as in bambino, for example) to neutron. When Pauli's proposal of the neutrino was discussed at a seminar in Italy, someone confused his neutral particle with Chadwick's particle. Fermi had to explain that Pauli's was the "little one."

It is certainly not necessary for the reader to master this lexicon in order to read on. For your convenience, I have provided Figure 11.4. At this point in our historical narrative, the photon and the graviton, the particles of light and of gravity, respectively, are in a class by themselves.

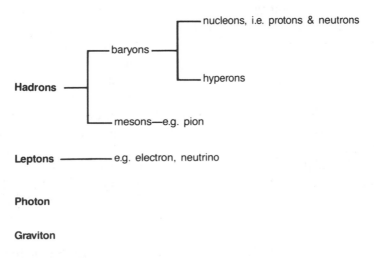

Figure 11.4.   Particles known circa 1960.

ORDERING THE CLUTTER

And so, to make a long story short, an alphabet soup of particles was known to physics by the late 1950s. Some organizational principle was needed to bring order to all the clutter.

Isospin was of enormous help. As I have remarked already, the power of symmetry is such that we know immediately that all hadrons must belong to isospin multiplets. It proved true experimentally. For example, three pions were discovered: π^+, π^0, and π^-. As the notation suggests, π^+ carries a positive electrical charge, π^0 is neutral, and π^- carries a negative electric charge. Thus, the three are distinguishable from each other by their electromagnetic properties, but as far as the strong interaction is concerned, they are identical. As expected, they each have almost the same mass. The mass of the π^+ and the π^- are measured to be 140 MeV, of the π^0 to be 135 MeV. The pions form a triplet, just as the two nucleons, the proton and the neutron, form a doublet. All hadrons were organized in this way into isospin multiplets.

As mentioned earlier, Noether's insight and strangeness conservation immediately implied that the strong interaction has a symmetry larger than isospin.

A CONVENTION OF TWINS AND TRIPLETS

Imagine that you are visiting a convention of twins and triplets. Here you meet a set of twins, there a set of triplets. Right away you notice a close family resemblance between the twins and the triplets. It turns out that they *do* come from the same family. By the late 1950s, particle physicists felt that Chadwick had introduced them to a convention of twins and triplets. They realized that subnuclear particles not only appear in doublets and triplets, but they are also as clearly related as members of a biological family.

Here is a summary of what was known about subnuclear particles around 1960. Altogether, eight baryons had been discovered. They were, first, the nucleon twins, our old friends the proton and the neutron. Then came the sigma and xi hyperons. Like the pion, the sigma comes in three varieties, positive, neutral, and negative, denoted by Σ^+, Σ^0, and Σ^-, respectively. Sigma hyperons belong to an isospin triplet; xi hyperons, as it turned out, to an isospin doublet. Finally, a hyperon, called the lambda and denoted by the Greek letter Λ, belongs to an isospin singlet all by itself.

Physicists summarize the situation using Figure 11.5. Each of the eight baryons is indicated by a dot plotted on a two-dimensional grid. The baryon corresponding to dots joined by horizontal lines belong to the same isospin multiplet. Baryons on the same horizontal level have the same strangeness. Thus, the nucleons have strangeness zero, Σ and Λ, −1, and Ξ, −2.

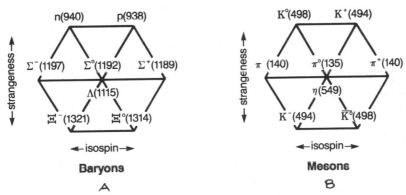

Figure 11.5. (A) The eight baryons, namely, the proton, the neutron, and their six cousins, are placed on a chart according to their isospin and strangeness. A geometrical figure called an *octet* is formed.

(B) The eight mesons also form an octet.

The situation regarding the mesons is quite similar. Physicists discovered eight of them. Besides the three pions, there are four mesons called *K* mesons, or kaons for short, belonging to two isospin doublets, and a meson called the eta, denoted by η, and belonging by itself to an isospin singlet. (Actually, only seven mesons were known at the time; the η was discovered later in 1961.)

In Figure 11.5, the mass of each baryon and meson is indicated in MeV units in parentheses after the letter denoting a given baryon or meson. Members of the same isospin multiplet have almost the same mass, as dictated by isospin symmetry. What is striking is that all eight baryons have roughly the same mass, to within 20 to 30 percent. They bear a definite family resemblance to each other.

The mesons also have roughly the same mass, clustering around a few hundred MeVs. The pions appear to be anomalously light. Nevertheless, one might argue that the eight mesons are related to first approximation.

Again, it is hardly necessary for the reader to memorize this zoo of subnuclear particles—any more than one needs to know by heart the dozen or so orders belonging to the class mammalia in the phylum chordata. The important thing to know is that the numerous species of mammals can be classified systematically, and that the principle of evolution ties all these groups together. Here too, the reader need only grasp that subnuclear particles appear to be related to each other.

A PSYCHOHISTORY OF PHYSICS

With all these look-alike particles, Nature was hinting that Her design possessed a symmetry larger than isospin. By the mid-1950s, the race was on to determine this larger symmetry.

The physics of an internal symmetry such as isospin was well understood: If the fundamental action is invariant under a group transformation, then there are quantum states that transform into each other and thus represent the multiplicative structure of the group. Physicists observe these quantum states as particles. As explained earlier, the particles transforming into each other must have the same mass if the symmetry is exact. To the extent that the masses of the observed baryons (and also observed mesons) are only approximately equal, physicists know that the symmetry, if there is one, must be even more approximate than isospin.

With hindsight, it would appear that the search for this larger symmetry should not have been too arduous. After all, Heisenberg had unveiled the internal world already, and mathematicians long ago had classified all possible transformation groups. Since there are eight baryons and eight mesons, physicists had only to find a group with an eight-dimensional representation, and it turns out that there are very few groups with such a representation.

One can only wish that it were so easy! Schoolchildren are taught a step-by-step schema of science in which one first gathers *all* the facts. Alas, in the real world, not all the needed facts are known, and not all known facts are true. For instance, in the late 1950s, experiments (now known to be wrong) showed that the Λ, under a parity transformation, behaved differently from the seven other baryons. It appeared that the baryon family included seven, not eight, members, and that the Λ was the odd man out. Many physicists were predisposed toward accepting this conclusion, since only seven mesons were known at the time (as mentioned earlier). Indeed, some of the most eminent physicists argued strongly that the baryon and meson families each contained seven members, and several physicists were misled, embarking on a fruitless study of groups with seven-dimensional representations.

Wrong experiments constitute one of the scourges of the theorist's life. Nowadays, experiments capable of throwing light on Nature's underlying design require heroic efforts, involving multinational teams. With technology stretched to the limit in search of minute signals, many experiments reach wrong conclusions, quite understandably. Increasingly, the ability to decide which experiment to trust is one of the necessary talents of the particle theorist.

Various psychosocial factors also hobbled the search for a higher symmetry. Those who had found isospin unattractive compared to exact spacetime symmetries were asked here to accept an even cruder symmetry. To an older generation whose nuclear physics characteristically involved energy measured at several MeVs, the mass splittings between the various baryons and between the mesons, in some cases amounting to several hundred MeVs, appeared enormous. To the Old Guard, the pions and kaons could not possibly be blood relatives. But what one generation calls enormous another calls tiny. My generation of physicists is used to experiments involving energies of several hundred thousand MeVs.

For these and other reasons, the search for the larger symmetry took a tortuous path. Indeed, several physicists essentially had arrived at what later turned out to be the correct symmetry group, only to be talked out of it by their senior colleagues.

THE TREACHERY OF NATURE

There were false starts, blind alleys. Early on, the Japanese physicist Shoichi Sakata tried the most obvious choice of a higher symmetry group: He went from Heisenberg's $SU(2)$ to $SU(3)$, the next group on the mathematicians' list. (Recall that the group $SU(2)$, in its defining representation, transforms two objects into each other. In general, the group $SU(N)$ transforms N objects into each other in its defining representation.)

In 1956, Sakata proposed that the proton, the neutron, and the lambda hyperon transform as a triplet in the defining representation of $SU(3)$. This proposal naturally generalizes Heisenberg's $SU(2)$, under which the proton and the neutron transform as a doublet. Tweedledum and Tweedledee were joined by Tweedledoo. But then the other five baryons do not fit. Attempts to shoehorn them in failed, and the group $SU(3)$ was apparently ruled out. It was puzzling that Nature did not progress in an orderly fashion from two nucleons to three, but jumped from two nucleons to eight baryons.

The breakthrough finally came in 1961. Murray Gell-Mann, the eminent physicist at Caltech, and Yuval Neéman, the military attaché in the Israeli embassy in London, working independently, concluded that the higher symmetry group—surprise, surprise—is none other than $SU(3)$. The catch is that the baryons belong not to the defining three-dimensional representation, but to an eight-dimensional representation of $SU(3)$. Nature had tried to trick us.

FROM TANKS TO $SU(3)$

How did a diplomat ever discover the higher symmetry of the strong attraction? As a youth, Yuval Neéman was interested in physics, but because of historical events, ended up in military service. He is said to have played an important role in the Israeli secret service. In 1957, at the age of thirty-two, Neéman, realizing that time was running out if he was ever to become a physicist, asked General Moshe Dayan, then Defense Chief of Staff, for a two-year leave to study physics at an Israeli university. Instead, Dayan assigned Neéman to the post of Defense Attaché in the Israeli embassy in London, a position which allowed him to pursue part-time study.

In London, Neéman went to see the distinguished Pakistani physicist Abdus Salam at the Imperial College, presenting his diplomatic credentials and a letter signed by Dayan. To his credit, the puzzled Salam took Neéman on. But in July 1958, the Middle East flared up once again and Neéman had to put physics aside. It is now known that, among certain other activities, Neéman had arranged for S-class submarines and Centurion tanks to be sent to

Israel. Finally, in May 1960, as a colonel on leave, Neéman enrolled for full-time study at age thirty-five, the oldest student in Salam's group. In his memoir, Neéman thanked his wife for accepting his change from diplomat to graduate student, and the consequent loss in income. Incidentally, Neéman has less time for physics these days. He founded his own political party in Israel several years ago and later became a member of the Israeli cabinet. The last time I saw him, we did not discuss physics; rather, he explained to me how to go about founding a party, in case I should ever want to do so.

THE EIGHTFOLD WAY TO NIRVANA

Once the symmetry group and the choice of representations were fixed upon, physicists could work out the experimental implications using group theory, as outlined in the preceding chapter. The symmetry forbids various subnuclear processes and allows others. The relative probability for various allowed processes to occur are determined. In this way, physicists, like lawmen moving into a frontier town, were able to impose some order on the subnuclear world.

To fit the baryons and mesons into the eight-dimensional representation, or octet for short, we have to check if the representation contains the correct isospin multiplets. In other words, since the symmetry $SU(3)$ contains isospin, we can ask what isospin representations are contained in the octet. With reference to the convention of twins and triplets, I can give the analog to this question. Suppose we meet eight siblings. We can ask how they separate into twins and triplets. They can separate, for instance, into two triplets and a set of twins ($8 \rightarrow 3 + 3 + 2$), or perhaps, two quartets ($8 \rightarrow 4 + 4$). Working out the $SU(3)$ group theory, one finds that the octet contains a triplet, two doublets, and a singlet ($8 \rightarrow 3 + 2 + 2 + 1$). But this corresponds precisely to observations: Of the eight baryons, the Σ hyperons form an isospin triplet, the nucleons and the Ξ hyperons form two doublets, and the Λ hyperon forms a singlet by itself. The same applies for mesons. The important point here is that finding an eight-dimensional representation is not enough; with the wrong group, the eight baryons may separate differently (according to $8 \rightarrow 3 + 3 + 2$, for example).

The exhilaration in finally determining $SU(3)$ as the higher symmetry may be likened to that felt by jigsaw-puzzle fans when the pieces suddenly all fit together. Gell-Mann was so pleased that he named the scheme the eightfold way, alluding to the eightfold path to Nirvana in Buddhism. (Incidentally, the eight are: right belief, right resolve, right speech, right conduct, right living, right effort, right contemplation, and right ecstasy.)

NO NEED TO LEAVE HOME

In spite of its successes, the eightfold way met with some skepticism, partly because of the psychosocial lack of readiness mentioned above. But the skeptics were silenced finally in 1964 by the dramatic discovery of a new particle.

Since the early 1950s, physicists had been discovering extremely short-lived particles, called resonances. By 1962, nine were known. The situation is summarized in Figure 11.6. As in Figure 11.5, the resonances within a given row are related by isospin, and the different rows are related by the eightfold way.

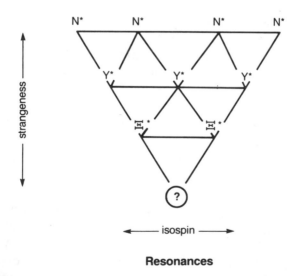

Resonances

Figure 11.6. The ten resonances, when arranged according to their isospin and strangeness, form a geometrical figure known as a *decuplet*. In 1962, only nine resonances were known. The eightfold way mandates the existence of a tenth resonance to fill the slot labeled with a question mark.

As I have emphasized, the power of symmetry considerations lies precisely in the fact that we do not have to know, in the present context, any details about the strong interaction physics of resonances to say immediately that resonances must belong to a representation of the symmetry group. Reasoning in this fashion, followers of the eightfold way rushed to see if $SU(3)$ has a nine-dimensional representation. It doesn't! But it does have a ten-dimensional representation. . . . Hmm.

Seizing on this group theoretic fact, Murray Gell-Mann predicted in 1962 that a hitherto unknown resonance, which he named the omega minus and wrote as Ω^-, must exist. Even more impressively, by using symmetry considerations Gell-Mann was able to predict all the relevant properties of Ω . It is the theoretical physics analogy to Sherlock Holmes's deducing the past experiences of a visitor just from a glance. A team of experimenters went out and looked. Sure enough, they found the Ω^-, with exactly the same properties as Gell-Mann predicted. For this and other fundamental contributions to physics, Gell-Mann was awarded the Nobel prize.

After the eighteenth-century French physicist Pierre de Maupertuis (whom we met earlier in connection with the action) survived an expedition to Lapland to verify Newton's theory on the flattening of the earth near the poles, Voltaire joked to him: *"Vous avez confirmé dans les lieux pleins d'ennui/ce que Newton connût sans sortir de chez lui."* (You have confirmed in places full of bothers/what Newton knew without leaving home.) In this case, the same may be jokingly said of the experimenters.

It never fails to amaze me how theorists are able, in the best cases, to predict the behavior of Nature by a few lines of reasoning. The purest products of the human intellect are often strikingly simple.

INSPIRED BY HAUTE CUISINE

Impressed by the power of symmetry, Gell-Mann forged ahead in search of more symmetries. How does a player get ahead in the symmetry game? Think of the historical precedents. Experimental observations spurred the development of isospin and the eightfold way. Lorentz invariance was born of Maxwell's theory. But in the early 1960s, with the observed pattern of strongly interacting particles accounted for already by the eightfold way, and without a theory of the strong interaction, Gell-Mann found neither of these precedents to be of any help. How could he proceed?

Boldly, Gell-Mann adopted a new strategy which he described as being inspired by a haute cuisine technique: "A piece of pheasant meat is cooked between two slices of veal, which are then discarded."

Gell-Mann imagined switching off all the interactions in the world. Without any interactions, particles would simply float about freely, paying no attention to each other. The theory describing this situation is known as a free theory and being trivial, it is dealt with in the first chapter of any physics textbook on field theory. Gell-Mann examined the action of free theory to determine its symmetries. He then proposed that some of these symmetries may still hold when the interactions are turned back on.

The procedure appeared manifestly absurd. Free theory most certainly does not describe our world. But Gell-Mann's attitude is that free theory, like the veal slices, will be thrown out once the symmetries—pheasant meat—have been extracted.

We might expect the symmetries so extracted to be rather crudely observed in the real world, if observed at all. The Old Guard, repelled by the crudity of isospin and the eightfold way, were due for yet another shock.

The proof is in the eating, of course. Did the veal help the pheasant? It turned out that considerable ingenuity was required to deduce experimental consequences from the symmetries extracted by Gell-Mann, but the consequences were all in agreement with observations.

The unexpected success of this blatantly absurd procedure —extracting symmetries relevant to Nature from a theory that definitely does not describe Nature—provided physicists with an important clue to the character of the strong interaction: The correct theory of strong interaction must have the same symmetries as free theory.

THREE QUARKS

After the eightfold way was established, physicists continued to puzzle over why Nature does not use the three-dimensional defining representation, the $SU(3)$ triplet. Did Nature simply want to trick Sakata?

In March 1963, Gell-Mann visited Columbia University. During lunch, Bob Serber, an eminent member of the Columbia faculty, asked Gell-Mann about the mysterious absence of the triplet. He replied that the particles transforming according to the triplet would have to have rather extraordinary properties; in particular, their electric charges would be smaller than that carried by the electron. No such particles had ever been seen.

The next morning, after mulling it over, Gell-Mann decided that such particles could have escaped detection. Subsequently, Gell-Mann, and independently, George Zweig, proposed that Nature does utilize the defining representation, and that the corresponding fundamental particles do exist. Gell-Mann referred to these three particles collectively as "quarks," naming them, individually, the up quark, the down quark, and the strange quark. Gell-Mann told me that he started out with a sound in mind: He wanted to call these triplet particles something like "kworks." One day, while idly leafing through James Joyce's *Finnegans Wake*, he came across the line "Three quarks for Muster Mark!," but he was disappointed that Joyce probably intended the word to rhyme with "mark" or "bark," rather than kwork. Then he realized that the book recounted the dream of a pub owner, and so he could imagine the three quarks to refer to three quarts. At this point, he went ahead and introduced quarks, pronounced kworks.

Using terminology introduced in Chapter 9, we can say that the three quarks furnish the defining representation of the group $SU(3)$. While Gell-Mann and Neéman spoke in 1961 of the defining representation in terms of three "abstract entities," in 1964 Gell-Mann and Zweig were able to speak of the defining representation realized by quarks. The abstract mental games of nineteenth-century physics are realized in the twentieth-century world of quarks!

The group theoretic process of gluing copies of the defining representations together, as discussed in Chapter 9, now can be visualized as actually combining quarks together. The group theory of $SU(3)$ naturally suggests that hadrons, the particles that interact strongly, are made of quarks and antiquarks, in the same way that an atom is made of electrons, and an atomic nucleus, in turn, is made of protons and neutrons. Put three quarks together, for example. We need only open a group theory book to find that

$3 \otimes 3 \otimes 3 = 1 \oplus 8 \oplus 8 \oplus 10$. In other words, gluing three copies of the defining representation together produces a one-dimensional representation, two eight-dimensional representations, and a ten-dimensional representation, thereby indicating that the eight baryons and the ten resonances can be constructed out of three quarks. Similarly, a study of the transformation properties of mesons suggests that a meson consists of a quark and an antiquark. The π^+, for example, is made of an up quark and an anti-down quark.

I may emphasize that the successes of the eightfold way stand, regardless of whether quarks exist or not. As has been stated repeatedly, the power of symmetry considerations is precisely that they do not depend on detailed dynamical knowledge.

The term quark, incidentally, has such a deliciously *recherché* ring to it that the physics community adopted it immediately, but Gell-Mann had to fight for the up, down, and strange quarks. In lingering remembrance of Sakata's theory (in which the fundamental entities were the proton, the neutron, and the lambda hyperon), most physicists, particularly on the East Coast, called the three quarks the P-quark, the N-quark, and the Λ-quark, written in capitalized script letters. I recall a senior physicist at Princeton exclaiming that people may be upside down and strange in California, but not here in New Jersey! This peevish disregard of Gell-Mann's will persisted well into the mid-1970s. For instance, the first papers on grand unified theories, to be discussed in Chapter 14, were written using this "East Coast" notation. I once gave a talk at a conference in Miami, and Gell-Mann was the session chairman. Every time I mentioned the P-quark, Gell-Mann would say, "up quark!" (I am reminded of a silly television commercial in which a man at breakfast battles a container of margarine. The man would say, "butter!," and the margarine would reply, "margarine!") This went on for a while, until finally I had to tell Gell-Mann to be quiet, reminding him that Miami is on the East Coast. Eventually, the community capitulated and grudgingly adopted the names, up, down, and strange.

A LAPSE OF FAITH

A multitude of hadrons has been constructed out of a few quarks: Reductionism triumphs once again. However, in the late

1950s and early 1960s, it was by no means clear that the reduction-ist approach would continue to work for the strong interaction.

Starting with Fermi and Yang, and continuing through Sakata, physicists tried to single out a few hadrons as special and then to construct the other hadrons out of these select few. Exper-imentally, however, one hadron looked pretty much like another, and the approach failed. A number of leading physicists, partly in reaction, felt that it was futile to ask what the hadrons were made of. Instead, they suggested constructing physics within a circular, logical framework. When asked what hadron A was made of, they would answer, hadron B and hadron C. And when asked what hadron B was made of, they would reply, hadrons A and D. And so on. It was hoped that this process would end, with hadron P, say. One would find that the hadrons A through P could all be thought of as made of each other. The number of hadrons would be fixed, without any reference to symmetry and group theory.

In this view, the world is the way it is because it is the way it is, a thought that an Eastern philosopher such as Chuang-tze might have had. The structure of the world is fixed by the necessity of mutual consistency between all phenomena. A school of physics based on this view flourished from the late 1950s until the early 1970s, and it acquired the suggestive name "bootstrap." One pic-tures the world pulling itself up by its bootstraps. Philosophers of science should have a field day studying this curious lapse of faith in reductionism.

QUARK CONFINEMENT

As a concrete model, the theory that hadrons are formed of quarks proves to be very useful: It helps physicists to visualize and to classify hadrons. For instance, the degree of strangeness of a hadron is measured just by the number of strange quarks it con-tains. One can account for the empirical rule that strange hadrons are more massive by supposing that the strange quark is more massive than the up and down quarks. The conservation of strangeness in the strong interaction may be accounted for if that interaction cannot change the character of the quark. In other words, a strange quark cannot be changed into a down quark (or an up), or vice versa, by the strong interaction. In a strong inter-action process, if a strange quark is created, it must be accompanied by an antistrange quark. As another example, consider the proton, which consists of two up quarks, a down quark, and the neutron, itself consisting of two down quarks and an up quark. Isospin is easily accounted for if the up and down quarks have approximately the same mass. (Ironically, Heisenberg's original thought that the mass difference between the proton and the neutron is due solely to electromagnetism is not believed to be entirely correct now. Part of this mass difference must be attributed to the mass difference between the up and down quarks.)

Given Gell-Mann's track record in predicting the Ω^-, many experimenters have rushed out to search for quarks, but all have come up empty-handed. Nevertheless, almost all particle physicists believe in quarks because numerous experiments have shown that hadrons behave as if they are made of quarks. In one of the most convincing experiments, electrons were accelerated down a mile-long tube at the Stanford Linear Accelerator Center to very high energies, then scattered off protons. The way the electrons were scattered showed quite clearly that they were bouncing off three particles inside each proton. To illustrate, let me borrow an analogy invented by George Gamow in a slightly different context: A customs official suspects that diamonds are being smuggled inside bales of cotton. He pulls out his pistol and starts firing at the bales. If some of the bullets ricochet and scatter, the inspector can be sure that something hard is hidden inside. By firing electrons at protons and watching them scatter, physicists are no less confident than the customs inspector that some particle resembling quarks live inside protons.

Yet the fact remains that no one has actually seen a quark. The situation has provoked philosophical musings on the meaning of existence. Again, another analogy: When I shake a baby rattle, I am absolutely positive that there are beads or something like beads hidden inside. By listening to the tone, and by varying the way I shake the rattle, I can deduce how hard the beads must be and, perhaps, even venture a guess on how many there are inside. My own position is such that, even if I am unable to crack the rattle open to actually see them, the beads surely exist. Similarly, physicists believe that quarks exist but that they are permanently confined inside hadrons.

That quarks are so confined is a shockingly unprecedented notion. When physicists suspected that matter is made up of atoms, it was easy enough to knock off an atom and isolate it for study. Similarly, one can readily knock electrons out of atoms, and protons and neutrons out of atomic nuclei. But, somehow, we just can't knock quarks out of hadrons.

The results of scattering electrons off protons indicate that the quarks inside the proton interact with each other in a rather peculiar way. If two quarks are close to each other, they behave as if they are almost free; in other words, as if they are barely interacting with each other. But if they get too far apart, a strong force suddenly tries to pull them together. To visualize this behavior, physicists think picturesquely of the quarks connected to each other by a string. When two quarks are close together, the string is slack, and the two quarks are not aware of each other. But if the two quarks try to move away from each other, the string suddenly becomes taut.

In classical physics, if we imagine pulling the two quarks

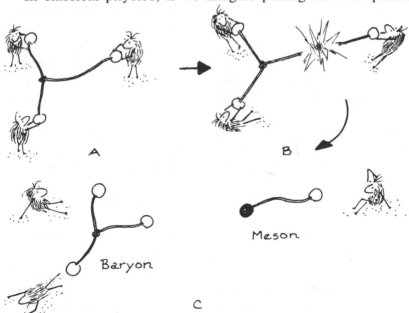

Figure 11.7. (A) Three hairy creatures pull hard on the three quarks in a baryon, trying to extract a quark.

(B) The string tying the three quarks together stretches and eventually breaks, thus releasing a wallop of energy symbolized as an explosion.

(C) The released energy is converted into a quark and an antiquark (represented as a black ball.) The three hairy creatures have only managed to knock off a meson. The conjecture that quarks are permanently confined means that the energy released by a breaking string is always converted into a quark and an antiquark.

apart hard enough, the string should break and the quarks would then be liberated from each other. But here, as we pull the quarks apart, we put energy into the string as a result of our exertion, in the same manner that stretching a rubber band puts energy into it. The energy in the string increases until it exceeds the energy corresponding to the masses of a quark and an antiquark, as indicated by Einstein's mass-energy relationship. At that point, the string itself has enough energy to create a quark and an antiquark. The string snaps, but out come a quark and antiquark. Even with the string broken we still have not managed to liberate a quark. The two broken ends of the string are attached, respectively, to a quark and to an antiquark. (See Figure 11.7.) We have only managed to knock off a meson.

The proposal of quarks illustrates the tortuous path the development of physics often takes. Consider the talk of the bootstrap versus reductionism. Actually, Gell-Mann was, and still is, a leading bootstrapper. I asked Gell-Mann about quarks versus the bootstrap. He recalled that it was all quite confusing. As a good bootstrapper, he believed that all observed hadrons are made up of each other. How then can the observed hadrons also be made out of quarks? Does this not mean that quarks are in turn made out of the observed hadrons? To escape from this insane conundrum, Gell-Mann finally decided that he could simply assert that the quark is not observable and hence not an "observable hadron." It is then not necessary to believe that quarks are made out of the observed hadrons. Quarks can be more fundamental. Whew! Thoughts along this line convinced Gell-Mann that quarks ought to be permanently confined.

In this chapter, physicists used symmetries to avoid dealing with the strong interaction. It is an end-run strategy, as indicated by the schema in Figure 11.2. But at some point one has to face the strong interaction. In the next chapter, I will discuss how physicists tamed the beast of strong interaction and explained the peculiar behavior of quarks.

12

The Revenge of Art

Zoe: Come and I will peel off.
Bloom: (Feeling his occiput dubiously with the unparalled
embarassment of a harassed pedlar gauging the symmetry of her peeled
pears.) Somebody would be dreadfully jealous if she knew.
—James Joyce, *Ulysses*

FLEAS IN MOZART'S COURT

When I think of the intellectual history of symmetry in physics, I like to picture two schools of thought, united in their devotion to symmetry but differing in their outlooks on the character of symmetry. On one side stand Einstein and his intellectual descendants. To them, symmetry is beauty incarnate, wedded to the geometry of spacetime. The symmetries known to Einstein— parity, rotation, Lorentz invariance, and general covariance—are exact and absolute, frozen in their perfection. On the other side stands Heisenberg with his isospin, shattering the aesthetic imperative of exact symmetry. Heisenberg's child is approximate and plays apart from spacetime. Unlike spacetime symmetries, isospin is respected only by the strong interaction.

The idea of an approximate symmetry appalled Einstein and his intellectual heirs. It seemed sacrilegious and oxymoronic to describe Nature as "approximately beautiful" and "almost perfect." The aesthetic sensibility of an entire generation was bruised. And no sooner had that generation of physicists gotten over the shock of isospin when along came the even more rudely approximate eightfold way. But, Heisenberg and his followers could point to the result—subnuclear clutter had been put in order. They marched, following the banner of symmetry, deep into the subnuclear forest, without having to determine the strong action first.

Using isospin, and later the eightfold way, they classified and made sense of the jumble of experimental observations.

By the 1960s, discussions on symmetry would focus almost entirely on the approximate symmetries of the subnuclear world. The absolute and exact symmetries known to Einstein continued to be held in the highest esteem, of course. Physicists all liked perfect symmetries; it was just that the fruitfulness of that particular notion appeared to have been exhausted, or rather, to have come to a gloriously climactic completion in the general theory of relativity.

While one would be deeply perturbed if spacetime were not ruled by exact symmetries, no pressing imperative demands that material particles must also obey exact symmetries. Thus, physicists reached a worldview in which Nature provides an exact spacetime, ruled by elegantly perfect symmetries, as an arena for a ragtag band of particles animated by a boisterously approximate code to play in. I picture a flea market held in the court of Mozart's Austria.

IN RETREAT BUT NOT IN DEFEAT

I have used a trick of the historian, in viewing the modern development of symmetry as resulting from the tension between two trends and outlooks. In this case, the outlooks on symmetry are not so much opposed as different. Furthermore, physicists who enjoy having Nature as the ultimate arbitrator, are not entrenched into warring camps as readily as other thinkers. Nevertheless, it is convenient to delineate sharply between these two views on how Nature "should" incorporate symmetry into Her design. I will continue, therefore, to speak of Einstein versus Heisenberg, of the exactness of art versus the crudity of pragmatism.

The devotees of the one true god of perfect symmetry were in retreat, but they were not defeated. While the banner of perfect symmetry was raised occasionally, the partisans lay low for almost forty years. Then they came roaring back in the early 1970s, eventually taking over all of fundamental physics. This vindication of art makes for a stirring story, a story which I will recount in this and the following chapters.

Figure 12.1.   A flea market in the court of Mozart's Austria.

COUNTERATTACK

The counterattack was mounted in 1954 in what can only be described as an epoch-making paper by Chen-ning Yang, whom we met before in connection with parity, and Robert Mills. It was unusual for a physics paper. It proposed a theory that did not appear to have anything whatsoever to do with the actual world.

Yang and Mills invented a new exact symmetry of dazzling mathematical beauty. The symmetry was not motivated by any experimental observation, as had been the case historically; rather, it was an intellectual construct based on aesthetics.

In a perfect illustration of Einstein's tenet that symmetry dictates design, Yang and Mills showed that their symmetry determines the form of the action completely. The situation was reminiscent of Einstein's determination of gravity theory, except that general covariance was motivated by Galileo's observation, while here the symmetry sprang out of purely intellectual considerations.

Alas, unlike the case of gravity, the action discovered by Yang and Mills did not fit the real world as perceived in the 1950s. The very presence of an exact symmetry implies collections of particles identical in their properties. No such collection had ever been seen.

Furthermore, one collection of particles, now known as gauge bosons, are forced by the symmetry to be massless, just like the photon. (More about gauge bosons later.) The point here is that to produce a massive particle, one has to supply an amount of energy at least equal to the mass of the particle. Thus, a massless particle is much more easily produced than a massive particle. Photons, for example, are easily produced. (That is why our world is filled with light.) It was a major embarrassment for Yang and Mills that the world was not filled with massless gauge bosons.

IN SEARCH OF A WORLD

Like a Pirandello character, the theory of Yang and Mills was in search of a world to describe. Their paper was not so much an offer to explain previously unexplained phenomena as a paean to the god of perfect symmetries. The paper seemed to say, "Look, here is the most beautiful theory that the human mind can dream up. If Nature does not choose to use this theory in Her underlying design, then we physicists could only be disappointed in Nature."

The exact symmetry proposed by Yang and Mills is now known by the rather forbidding-sounding name of "non-abelian gauge symmetry." The theory dictated by this symmetry is known either as a non–abelian gauge theory or a Yang–Mills theory.

When Yang–Mills theory first came out, the community of theoretical physicists agreed that it was indeed beautiful, but no one, not even Yang and Mills, had the foggiest idea what it was good for. Most physicists simply mumbled that it is too bad we do not live in a non–abelian gauge world, shrugged, and went on with whatever they were doing. The theory thus lay dormant.

When I was in graduate school in the late 1960s, non-abelian theory was not taught. Physicists studying the strong interaction focused on phenomenological theories, which seek to account for the actual details of observation. The pragmatic philosophy embodied in these studies is diametrically opposed to the aesthetic philosophy so deeply felt by Einstein, and some of these theories, though successful in making sense of the data, were in fact brutally ugly.

In the rest of this chapter, I will first explain non–abelian gauge symmetry, then tell the exciting story of how physicists came to realize that Nature worships the same god worshiped by Einstein's intellectual disciples—and that the fabric of Nature is designed around non–abelian gauge symmetry.

STRANDS INTO TAPESTRY

I have presented the development of twentieth-century physics as an intellectual history. Out of this history of ideas, elements emerged from the work of Einstein, Noether, and Heisenberg to fuse into the concept of non–abelian gauge symmetry.

From Einstein's work on gravity came the notion of local transformation. Recall that Einstein's strategy for dealing with an arbitrary gravitational field was to divide spacetime into smaller and smaller regions, so that, within each region, the gravitational field is constant to an ever-increasing accuracy. In this way, one ends up with the idea that the coordinate transformations one has to perform to mock up the gravitational field vary from point to point. A symmetry involving transformations that vary from point to point is said to be local.

On the other hand, a symmetry involving transformations that do not vary from point to point is said to be global. For a global symmetry, everyone in the universe would have to perform exactly the same transformation in order to leave the structure of physical reality invariant.

Isospin invariance provides an example of a global symmetry. Heisenberg postulated that strong interaction physics is invariant under transformations turning the proton into the neutron, and vice versa. The strong interaction cannot distinguish between the proton and the neutron. In other words, in the approximation in which one neglects the three feebler interactions, it does not matter which of the two nucleons one calls the proton, which the neutron. But once we decide which one we call the proton, we have to stick to that same choice throughout the universe. Put another way, if we perform an isospin transformation rotating the proton into the neutron, we have to perform the same rotation everywhere throughout the universe in order to leave the action invariant.

Locality in symmetry transformation is one of those concepts that seems utterly natural, once someone enunciates it. If I perform a symmetry transformation on earth, another physicist on the dark side of the moon, or for that matter three galaxies over, ought to be able to perform some other symmetry transformation. In contrast, many theorists feel vaguely uneasy with the notion of global symmetry; in fact, it was largely distaste for global symmetries that drove Yang and Mills to propose their theory.

Another all-pervasive strand in modern thought is the deep connection between conservation and symmetry. Recall that Noether's insight prompted a search for the symmetry responsible for electric charge conservation, a search taken up by the physicist Hermann Weyl. The symmetry in question turns out to be rather peculiar and fairly abstract. For our purposes, it is only necessary to know that such a transformation exists.

Inspired by Einstein, Weyl decided to demand that the symmetry responsible for electric charge conservation be local. He discovered, to his surprise, that this requirement leads to a striking consequence.

I mentioned earlier that theoretical physicists dream of scribbling down the action of the world on a cocktail napkin. The action certainly would include a term describing the electromagnetic interaction. We know, after all, that the electromagnetic interaction exists. To understand Weyl's work, imagine a theoretical physicist who had a lapse of professional competence and had forgotten to include the electromagnetic field in the action. Now Weyl comes along and looks at the napkin. "Hmm, let me check this action to see if the symmetry responsible for electric charge conservation is local." (Strictly speaking, I should not use the phrase "electric charge" in the absence of an electromagnetic field. But call it what you will, "electron number," for example; it does not affect the argument.) In fact, in the action without electromagnetism, the symmetry in question would not be local. It is global. Most remarkably, Weyl showed that, conversely, if one requires this symmetry to be local, one is *forced* to include the electromagnetic field—and hence, light.

This was an astonishing discovery. Physicists have worked to understand the properties of light, but they always thought that the question of why there is light was beyond their ken. That "why" has now been replaced by the question of why Weyl's symmetry has to be local. One why has been replaced by a more profound why. With this we have come a little closer to knowing His thoughts. The question of why symmetries should be local can now be debated on aesthetic grounds. How did the mind of the Creator work when He designed our cosmos? Did He say, "Let there be light!" or did He say, "Let symmetries be local!"?

The power of local symmetry was already manifest in Einstein's theory of gravity, which is, after all, the original example of a locally symmetric theory. In the same way that Weyl's local symmetry forced the photon on Weyl, local coordinate invariance forced the graviton on Einstein. Suppose we never heard of gravity but decided to require that the action of the world be invariant under local general coordinate transformations. We would find that we have to invent gravity.

Weyl named his symmetry gauge symmetry. The term "gauge" comes from low Latin *gaugia,* referring to the standard size of casks, and this sense is retained in such modern usage as "railroad gauge" and "gauged skirt." Curiously, the word entered the permanent vocabulary of physics only because Weyl made a serious but justifiable mistake. We now know that the symmetry responsible for electric charge conservation is described by transformations involving the quantum probability amplitude. Weyl was working before the advent of quantum physics, so he, like everyone else, never dreamed of probability amplitudes. Instead, inspired by the geometric flavor of Einstein's work, Weyl proposed a transformation in which one changes the physical distance between spacetime points. Weyl was reminded of the distance, or gauge, between two rails—hence the name for his symmetry. He showed Einstein his theory, but they were both deeply disappointed that it failed to describe electromagnetism. When the quantum era began, Weyl's theory was quickly repaired. Meanwhile, the term gauge symmetry, although a misnomer, remained. (Incidentally, physicists still do not know whether Weyl's original symmetry is relevant to the world.)

In summary, the action of the world, including electromagnetism, possesses a local symmetry called gauge symmetry. In traditional textbooks, students were presented with the action describing electromagnetism and told, "Behold! There is a local symmetry." Nowadays, fundamental physicists, following Einstein, prefer to reverse the logic and say that the local symmetry fixes the form of the action. Symmetry dictates design.

The story of gauge symmetry illustrates how physics becomes simpler. Recall that, as a student, I had to memorize the four Maxwell equations. Later, I needed only to remember an equation describing how the electromagnetic field varies in space-time, an equation almost as easy to remember as the shape of the circle. Now, I just say "gauge symmetry" and electromagnetism is determined.

Finally, Heisenberg came along and opened up an entirely new internal world for theoretical physicists to cavort in. But as we have explained, the symmetry he proposed for his internal world is approximate and rather ugly, and the transformation involved is global.

Yang and Mills fused these different strands. They took Heisenberg's notion of an internal symmetry, but they insisted that it be exact. Then they made this exact symmetry local, a notion born of Einstein via Weyl. The result was called a non–abelian gauge symmetry.

QUANTUM FIELD THEORY

I should mention here that fundamental physics has been formulated in the language of quantum field theory since the 1930s. Before the quantum era, there was a dichotomy between particles and fields in our description of Nature. Particles such as the electron and the proton produce electromagnetic and gravitational fields. These fields act, in turn, on particles and affect their motion. To specify the dynamics of a particle, physicists have to give the location of the particle at a given time. In contrast, to specify the dynamics of a field, physicists have to give a bunch of numbers at each point in space at a given time. In the case of the electromagnetic field, the numbers are simply those characterizing the magnitude and direction of the electric and magnetic forces which a test probe would detect at that point. Thus one can picture a field pervading spacetime.

With the dawning of the quantum era, this dichotomy between particles and fields was removed. Particles now are described by the probability amplitude waves governing their motion, and these waves are specified at each point in space and in time. In other words, particles in the quantum world also are described by fields. Faraday's concept of field has taken over all of physics.

In a quantum field theory, the action is constructed out of fields combined in such a way that the action will satisfy whatever symmetries are desired. For example, to construct a quantum field theory describing the interaction of electrons and photons, simply combine the electron field and the photon field (that is, the electromagnetic field) into an action which satisfies Lorentz symmetry and gauge symmetry. It is really quite simple once one learns how to do it.

A CONSTRUCTION MANUAL

To explain Yang–Mills theory further, let me give the recipe for constructing the action. Choose any one of the groups known to mathematicians and postulate an internal symmetry based on that group. (The group does not have to be $SU(2)$, as is the case in Heisenberg's work.) Insist that the action be invariant under local symmetry transformation. The action is then Yang–Mills. (It is understood that the action is also invariant under the established symmetries of spacetime, such as Lorentz symmetry.) Incidentally, the electromagnetic action described by Weyl is just a special case of a Yang–Mills action.

In Weyl's discussion, the local symmetry demands the presence of the electromagnetic field, with the associated massless photon. In a non–abelian gauge theory, the local symmetry demands the presence of a certain number of fields, each associated with a massless particle. If you follow the recipe given above for constructing the Yang–Mills action, you will find that, try as you may, you will not be able to make the action locally invariant unless you add these extra fields. They were forced on Yang and Mills, in the same way that the photon field was forced on Weyl.

We saw how a global symmetry such as the eightfold way

demands the presence of a tenth resonance, given that there are nine resonances. To draw a crude analogy, if an architect's client insists on a building with pentagonal symmetry and the design already includes four columns, the architect will have to put in a fifth column. Local symmetry is even more demanding than global symmetry: Not only does it require new particles, it also specifies that these particles must be massless.

Let me indicate roughly why this is so. Local symmetry allows different transformations to be performed in different regions of spacetime. To be specific, suppose for a moment that isospin is a local symmetry. I choose to call one of the nucleons the proton. But the man on the other side of the moon may choose to call what I call the proton the neutron. To transmit information about my choice to my lunar friend, a long-range field is needed. Recall that massless particles are associated with long-range fields. Thus, the appearance of massless fields in gauge theories is not entirely surprising. Incidentally, this is a hand-waving explanation; in physics, an argument that is not totally convincing is so known because the one presenting it typically has to gesticulate a lot.

The massless particles in Yang–Mills theory are called "gauge bosons." Once we decide on a group, the number of gauge bosons is determined completely. As I mentioned, electromagnetic theory is a special case of Yang–Mills theory. It has only one gauge boson—the photon.

GAUGE DYNAMICS

While gauge bosons are required by gauge symmetry, the particles that interact with the gauge bosons are not. Theoretical physicists can put them in as they see fit. Each particle is associated with a field. Under the internal symmetry, these fields transform into each other, thus furnishing a representation of the group. Recall that the group transformations scramble together a bunch of entities. The number of entities in a given representation is completely determined by group theory. We can choose any representation we like, but once the choice is made the number of fields is completely fixed.

The basic dynamics of a non–abelian gauge theory can now be described quite simply. When a particle emits or absorbs a gauge boson, it changes into another particle. In other words, a gauge field transforms particles into each other.

The preceding description may remind the reader of our discussion of group representations. Mathematicians have long thought of group transformations acting on the abstract entities in a representation. Here, these abstract entities are replaced not by kittens or apples, but by particles and their associated fields. The particles are actually transformed into each other by the gauge bosons. What a thrill to see the abstract musings of mathematicians actually realized in the physical world!

But I am getting ahead of myself. I have yet to tell the story of how Yang–Mills theory found a world to describe.

PHYSICISTS AS DESPERADOES

The story begins back in the 1950s, with the long-standing difficulty of constructing a theory of strong interaction. The perturbation approach had failed utterly in dealing with the strong interaction. It appeared as if we would never be able to determine in detail the precise structure of the strong interaction. The best we could do was explore the symmetry of the strong interaction. Perhaps I should emphasize that physics did not fail, but rather the computational methods of physics failed. The strong interaction is too strong and that is that. In desperation, some physicists in the early 1960s even advocated abandoning reductionism in studying the strong interaction. As I already mentioned, this proved to be a glitch in the intellectual history of physics, and reductionism was eventually vindicated. How in the world did physicists ever manage to tame the strong interaction then?

Since I participated to some extent in the astonishing turnabout in our view of the strong interaction, I will tell the story from a personal point of view. What follows should not be taken by any means to be a history of the modern theory of the strong interaction. Rather, it is my first-person experience of history. Just as an actor who went offstage to change does not know firsthand what transpired onstage in his absence, I cannot possibly describe how other physicists view this period.

In the spring of 1970, as I was about to receive my degree in physics, the physicist Roman Jackiw asked me if I would like to spend some time that summer in Aspen, Colorado. Every summer, theoretical physicists from all over the world gather in Aspen to exchange ideas amid the majestic setting of the Colorado Rockies. I naturally jumped at the opportunity. When I arrived, I learned that, as a greenhorn rookie physicist, I had been assigned to live in a basement, but I was pleasantly surprised when I discovered that I was to share it with Ken Wilson, a physicist well known for his profundity of thought. I telephoned my then-future wife that I could now indulge in the elemental emotion of hero worship to the fullest. I often read in the sports page about a rookie football player describing his emotion on finding himself playing side by side with a great he had admired ever since junior high. Well, a physicist feels pretty much the same way for the first few years after receiving his degree.

The basement was not subdivided into rooms, so I got to know Ken Wilson rather well. We ate dinner together every night and I learned a fabulous amount of physics from him. At that time, Ken Wilson had just finished a massive piece of work that would later win him the Nobel prize. He asked me to go over the manuscript and to point out the passages that were unclear to me. It is an intriguing feature of the human mind that among truly deep thinkers, some are lucid expositors, others almost impossible to understand. I must confess that I had to struggle terribly to understand Ken.

LOOKING AT THE WORLD

Wilson was concerned with how we describe the world. The reader is familiar, of course, with the fact that the world looks quite different when examined on different length scales. Increase the resolution of the microscope and what appeared as haze crystallizes into detailed structure. In examples drawn from everyday life, our perceptions of the world, with a given resolution, do not tell us very much about what we will see with a finer resolution. However, the logical structure of quantum field theory is so intricate that it can relate a description on one length scale to a description on another. Given a description of the world, physicists can actually say something about how the world would look if seen with finer resolution. The essence of Wilson's work deals with how much the logical structure of quantum field theory allows us to say.

In the preceding chapter, I said that each of the four fundamental interactions is characterized by a coupling constant that measures the strength of the interaction. (In quantum physics, the probability that two particles will interact determines the strength of the interaction.) Physicists in 1970 were used to thinking of coupling constants as constants, a notion built into the term. But let us rely on operationalism and think how one of my experimental colleagues would actually go about determining the coupling constant of the electromagnetic interaction. Well, he would collide two electrons together, for example, and by repeating the experiment many times, determine the probability that the electrons would actually interact. That probability, essentially, defines the electromagnetic coupling constant. This operational definition makes clear that the coupling constant will depend on the energy with which the experimenter collides the two electrons. Another experimenter repeating the measurement at a different energy will extract a different coupling constant. The so-called coupling constant is not a constant at all, but varies with the energy scale at which it is measured. So, instead of coupling constant, I will henceforth use the more appropriate term coupling strength, or simply coupling.

Recall that in quantum physics the wavelength of any particle that we use as a probe decreases as the energy of the particle is increased. Thus, to examine Nature with a finer resolution, physicists simply collide particles at higher energies. This discussion tells us that as we look at Nature with different resolutions, the coupling strengths of the various interactions will vary.

One aspect of Wilson's work deals precisely with this variation. Interestingly, the internal logic of quantum field theory allows us to determine how the coupling strength varies with energy, at least in principle. Speaking picturesquely, physicists say that the coupling strength "moves" as one changes the energy at which one measures the coupling strength. It turns out that the coupling strength moves extremely slowly with energy. Over the entire range of energy that has been studied from the beginning of physics until the late 1960s, the electromagnetic coupling strength has moved only by a minute amount. This explains why the coupling strength had always been thought to be constant.

MOMENTS OF WEAKNESS

In the fall of 1970, after my encounter with Ken Wilson, I went to the Institute for Advanced Study in Princeton as a postdoctoral fellow. One of my friends there convinced me that having come from Harvard, where reductionism continued to be held in high regard, I was deficient in my education. I thought I had better try nonreductionism for a while. I did not return to the idea of moving coupling strengths until 1971. In the summer of 1971, my thesis adviser, Sidney Coleman, who is well known as a gifted expositor, gave some lectures on moving coupling strengths and using a formulation given by Curt Callan and Kurt Symanzik, managed to make the subject crystal clear. I studied Coleman's lectures in detail, but was discouraged by his rather pessimistic conclusions. Once again, the only calculational procedure available was based on the perturbation method. Consequently, one could determine the movement of coupling strengths only when the coupling was small, so physicists were just as helpless as before in dealing with the strong interaction. The strong coupling would move, but like Longfellow's arrow, to physicists knew not where.

One day, in the spring of 1972, as I lay on the couch rereading Coleman's lectures, the thought occurred to me that perhaps Nature is kinder than we think. My idea was that as we look at the strong interaction with ever-increasing energy, the strong interaction coupling may perhaps move to zero. If so, the strong interaction could be tamed after all. Mr. Macho-Too-Strong may yet have his moments of weakness!

But in order to see if this actually occurs, I was faced with the same old problem of the strong interaction coupling, being strong, preventing me from doing any meaningful calculations. I needed another strategy. I decided to ask myself the following question instead. Suppose, at some energy scale, the strong coupling becomes quite weak: Would it tend to become stronger or weaker as we move up in energy? Now, this is a calculation I could do using the perturbation approach, since by supposition the strong coupling had already become weak. Let me make a rough analogy. While I am not a social economist, I can readily imagine that it is easier to predict whether a given poor family will get poorer than to predict whether a middle-class family will get richer or poorer.

This supposition appears quite strange at first glance, and the senior physicists to whom I had mentioned it reacted rather negatively. The strong coupling has always been strong. To ask what would happen if the strong coupling were weak may be vaguely reminiscent of the old joke in which the raconteur asks what would his aunt give him for Christmas if she happened to be a man.

Nevertheless, I thought the idea was well worth pursuing. If the calculation shows that the coupling, once weak, will become weaker, then physicists at least can hope that at some point the strong coupling will tend toward zero. I said earlier that a theory with zero coupling is said to be "free"—the particles in it would be free to move about independently of the other particles. A theory whose coupling strength moves toward zero as the theory is examined at ever higher energies is now called an "asymptotically free" theory.

The reader may get the impression that the search for asymptotic freedom was motivated purely by wishful thinking. We would like the strong interaction to become weaker at higher energies so that we can deal with it. This is almost, but not exactly, true. Around 1970, there was already a hint, at least in hindsight, that the strong interaction may get weaker at high energies. Experimenters had scattered very energetic electrons off protons. In the quark picture, the electron gives one of the quarks inside the proton a good kick. I have already stated in the preceding chapter that the experimental results indicate that when the quark kicked by the electron zips by the other quarks, it barely interacts with the other quarks. Asymptotic freedom was able to explain this phenomenon naturally.

THE SEARCH FOR FREEDOM

The attentive reader may be puzzled as to how I proposed to do the calculation since I did not know, and neither did anyone else, what the theory of strong interaction might be. In order to proceed, I had the idea of looking at all possible types of theories. Those theories that are asymptotically free would then present themselves as attractive candidates for a theory of the strong interaction. In searching for asymptotic freedom, one may end up finding the theory of strong interaction as well!

And thus I embarked on a journey of the mind, looking at each theory in turn, in search of asymptotic freedom. It was quite a treasure hunt, made all the more exciting since I did not know beforehand whether there was any treasure to be found.

Unfortunately, none of the theories I looked at were asymptotically free. For example, the electromagnetic coupling increases as the relevant energy of the electromagnetic process increases. (It will turn out later that this fact is of crucial importance for a unified understanding of the universe.) I was rather disappointed.

In the fall of 1972, I joined the faculty of the Rockefeller University in New York. It was an exciting time for me personally: I was newly married, and my wife and I were released from Princeton into the big city, with all its attractions. It was also an extraordinarily exciting time in particle physics. Physicists started to believe that the electromagnetic and weak interactions could be unified into a Yang–Mills theory. (I will go into *that* story in the next chapter.) The collective consciousness of fundamental physicists was awakening to the Yang–Mills theory. Since the proposed unification of the electromagnetic and weak interactions has not yet been confirmed experimentally, many theorists were seized by a frenzy to construct competing theories of these interactions. One hardly knew whether to work on the strong or the electromagnetic and weak interactions, or, in my case, to sample the New York nightlife.

I decided to do a little of each. Here I will tell the reader only of my search for freedom. Up to that point in my search, I had examined all the theories I had learned about in graduate school. But I had not yet examined this Yang–Mills theory, which everyone was now talking about in connection with the electromagnetic and weak interactions, so I decided to do that next.

In the winter of 1972–73, I heard, as by thunder, the electrifying news that freedom was found. David Gross and his graduate student, Frank Wilczek, at Princeton University, and, independently, David Politzer, a graduate student of Sidney Coleman's at Harvard University, found that Yang–Mills theory is asymptotically free. The news was stunning—we may finally have a handle on the strong interaction! Shortly thereafter, I saw Ken Wilson and Kurt Symanzik in Philadelphia at a conference on another subject, and I remember how excitedly we discussed the news. Symanzik and I took the same train coming back, and we spent the time chatting about the movement of couplings and asymptotic freedom.

When I got back to New York, I tried to work out the experimental consequences of asymptotic freedom. At that time, experimenters were studying the annihilation of an electron and a positron—an antielectron—into strongly interacting particles. It had long been known that matter and antimatter annihilate into a burst of energy out of which particles materialize. I picked this process because, in contrast to what the reader may think, it has an extremely simple theoretical description. I showed that with asymptotic freedom, the probability of electron-positron annihilation into strongly interacting particles should decrease in a definite way as the energies of the colliding electron and positron increase. (Similar work was done independently by Tom Appelquist and Howard Georgi, and by David Gross and Frank Wilczek.)

I then called up an experimentalist, and to my utter disappointment, I learned that the probability in question in fact increases as the energies of annihilation increase. So much for our dreams of freedom, then! The beauty of the theory, however, was so beguiling that I published my work, anyway. Later, those experiments were found to be wrong.

Gross and Wilczek, and somewhat later, Georgi and Politzer, tackled the theoretically more difficult problem of electron scattering off protons in the experiment I mentioned earlier. The detailed agreement between their theoretical calculations and the experiments established that Nature really does enjoy asymptotic freedom.

Meanwhile, David Gross had offered me a faculty position at Princeton, so I returned to the tranquillity of pastoral New Jersey in the fall of 1973. (Incidentally, I had been offered a similar position the year before, but due to a curious twist of fate, had gone to New York instead.) At Princeton, I worked with Sam Treiman and Frank Wilczek, studying the detailed effects of asymptotic freedom on the scattering of highly energetic neutrinos off protons.

Asymptotic freedom has been firmly established by now; physicists have tamed the strong interaction. The discovery of asymptotic freedom by Gross, Wilczek, and Politzer ranks as a great triumph of theoretical physics.

The story of asymptotic freedom illustrates the notion prevalent among the disciples of Einstein that a theory dedicated to beauty will naturally have wonderful properties. Furthermore, one can show that Yang–Mills theory is the only asymptotically free theory in our spacetime.

Good. Yang–Mills theory is asymptotically free. But how can we use it to describe the strong interaction? The theory is constructed on an exact symmetry, but the known symmetries of strong interaction are manifestly approximate. To describe the world, the theory must go in search of an exact symmetry to hang its hat on. To explain how this symmetry was found, I have to go back to 1967.

ALL GOOD THINGS COME IN THREES

By 1967, most physicists were ready to believe in quarks. To put the idea of quarks to a test, theorists would have liked to calculate a physically measurable quantity involving quarks. But since quarks interact strongly, no one knew how to proceed. Then in 1967, in an ingenious work, Steve Adler and Bill Bardeen, and also James Bell and Roman Jackiw, showed that of the bewildering multitude of processes involving the strong interaction one quantity could be calculated, namely the lifetime of the electrically neutral pion. Here, a variety of factors conspired so that the unknown effects of the strong interaction all cancelled out. Physicists were overjoyed, but also puzzled. When the calculation was done, the amplitude for the neutral pion to disintegrate was found to be off by a factor of three from the value deduced from the experimental measurement.

Perplexed, physicists soon realized, however, that this discrepancy could be explained if there were three times as many quarks as they thought there were! In the previous chapter, the reader was introduced to the up quark, the down quark, the strange quark, and so on. Suddenly, it appeared that each of these quarks came in three copies each. Gell-Mann introduced the picturesque term "color" to describe this bizarre triplication. Each quark supposedly comes in three colors—red, yellow, and blue, say. Thus, there is a red up quark, a yellow up quark, a blue up quark, a red down quark, et cetera. (The reader should understand that the term color is used metaphorically.) More evidence soon surfaced supporting this triplication of quarks. The situation was

most disconcerting. Why should Nature be so extravagantly florid?
Poetry should be lean and sparse.

COLOR SYMMETRY

Once physicists contemplated the possibility that a Yang–
Mills theory could describe the strong interaction, Nature's pur-
pose in triplicating the quarks became clear at once. An up quark
is certainly different from a down quark; they have different
masses, for example. But suppose the red up quark, the yellow up
quark, and the blue up quark all have precisely the same mass, and
similarly for the down quark, the strange quark, and any other
kind of quark. Then we have the exact symmetry we need in order
to have a Yang–Mills theory! What appeared as florid extrava-
gance in the late 1960s turned out to be eloquent purposefulness
on Nature's part.

The symmetry involves transforming a quark of one color
into the same kind of quark but of a different color. The fact that
differently colored quarks have exactly the same mass explains
why it took physicists a while to catch on to this bizarre triplication. After all, we may nod hello to a neighbor for years before
discovering that she is a member of an identical triplet—and that
we have been seeing three different persons all along! Dramatists,
including Shakespeare, have used this device to complicate their
plots or trick the audience. Physicists also felt that Nature had
played a trick on them.

Incidentally, to avoid having to say "the same kind of
quark" all of the time, physicists find it convenient to say that each
quark carries two attributes: flavor and color. Thus far, I have
introduced three different flavors: up, down, and strange. Each of
these three flavors comes in three colors, making nine quarks in
all. Imagine an ice-cream store whose customers are so choosy
that they demand not only thirty-one flavors, but also that each
flavor be served in their favorite color. The price of a cone may
vary from flavor to flavor, but, for a given flavor, the price is the
same regardless of color. Artificial coloring is cheap. Amusingly,
the world of quarks appears to be set up in pretty much the same
way. The mass of a quark is the same regardless of its color, but
varies from flavor to flavor.

Figure 12.2. A quark parlor: Six flavors are available; each flavor comes in three colors. Up is the least expensive flavor; top, the most expensive. The price is the same for each regardless of the color.

In this language, the Yang–Mills symmetry changes the color carried by a quark, but not the flavor.

Under the symmetry, three different colors are transformed into each other. The relevant group, therefore, is just $SU(3)$, sometimes referred to as color $SU(3)$, to distinguish it from the $SU(3)$ of the eightfold way.

Gell-Mann named the eight gauge bosons contained in the theory "gluons," since they are supposed to glue the quarks into hadrons. The force between quarks is mediated by the exchange of gluons in the same way that the electromagnetic force between charged particles is mediated by the exchange of photons. This theory of the strong interaction is now called "quantum chromodynamics."

FREEDOM AND SLAVERY

Asymptotic freedom means that the strong force becomes weak when two quarks collide at high energies, or equivalently, according to quantum physics, when they get very close to each other. The quarks become barely aware of each other; they are free from each other's influence. (This explains the terminology of asymptotic freedom—the closer quarks get to each other, the freer they "feel.") Thus, one can use the perturbation method to calculate those strong interaction processes in which quarks get close to each other. This marvelous situation is, of course, what motivated the search for asymptotic freedom in the first place. Since the results of these calculations agree with observations, quantum chromodynamics now is almost universally accepted as the correct theory of the strong interaction.

The flip side of freedom is slavery, however: As the two quarks move away from each other, the coupling strength moves away from zero. It is generally believed today that the coupling strength becomes stronger and stronger, and prevents the two

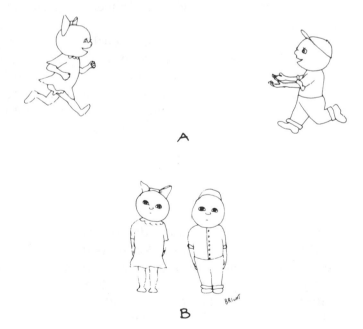

Figure 12.3. Quarks are like some lovers: When they are far apart (A), they want each other, but when they are close together (B), they barely acknowledge each other's presence.

quarks from ever separating from each other. This phenomenon is known as "infrared slavery." (The adjective infrared is of purely historic origin and does not concern us here.)

Infrared slavery neatly explains the fact that no one has ever seen a quark: Quarks are confined inside hadrons.

Quarks behave like some lovers. When they are far apart, they pine for each other with zeal, vowing that no one can ever separate them. But once they get close to each other, they replace their ardor with indifference and barely communicate with each other.

The interaction between quarks is quite contrary to our intuition about what interactions between particles should be like, an intuition built up from studies of other interactions. The electromagnetic interaction between two electrons, for example, decreases as the electrons move apart. From playing with magnets, we know that as we separate two magnets, the interaction between them gets progressively weaker. But as you pull two quarks apart, the interaction between them gets progressively stronger!

According to quantum chromodynamics, if quarks are enslaved and confined, so are gluons. Indeed, experimenters have never seen a gluon.

That quantum chromodynamics enslaves and confines has never been proven, precisely because physicists have been unable to deal with strong forces in the first place. But theorists have compiled a fair amount of circumstantial evidence supporting the notion. Some theorists feel that the enslavement, is not absolute; perhaps, if one pulls at the lovers hard enough, they could be separated.

APPEARANCE AND REALITY

Quantum chromodynamics drastically revises our understanding of the strong interaction. We thought that the strong force was mediated by the pion, but in fact, at a deeper level, the strong force between quarks is mediated by gluons. The gluons glue quarks into nucleons, pions, and other hadrons. The force between two nucleons caused by the exchange of pions between them is merely the phenomenological manifestation of a deeper reality. As an analogy, we might think of two complicated biochemical molecules coming close to each other. A piece of one molecule may break off and attach itself to the other molecule. Certain biological processes work in exactly this fashion. But this interaction between the molecules is clearly just a phenomenological manifestation of the underlying electromagnetic interaction between the electrons and the atomic nuclei. As fundamental physicists once moved from molecules to atoms, from atoms to nuclei, from nuclei to hadrons, they now move from hadrons to quarks.

As a theory, quantum chromodynamics has no precedent in the history of physics. The present understanding is that the basic constructs in the theory, quarks and gluons, cannot be observed directly, even in principle.

THE REVENGE IS COMPLETE

Isospin and the eightfold way, the strong interaction symmetries of old, are now revealed to be incidental symmetries that tell us almost nothing about the underlying essence of the strong interaction. In the quark picture, isospin transformations change the up quark and the down quark into each other, while the transformations of the eightfold way change the up, the down, and the strange quarks into each other. They change flavor, but not color. They are symmetries only to the extent that these three quarks have approximately equal masses. Our understanding at present is that the masses of differently flavored quarks are not controlled by the strong interaction, and that there is no intrinsic reason for them to be equal. In fact, as I will discuss in a later chapter, quarks with other flavors have been discovered in the last ten years, and their masses are quite different from the up, the down, and the strange. On the other hand, quarks with the same flavor but with different colors always have exactly the same mass.

The essence of the strong interaction is controlled by the exact symmetry of Yang and Mills, a symmetry born out of Einstein and Weyl and touched with the logic of geometry, rather than by the approximate symmetries of isospin and the eightfold way. The revenge is complete.

LEARNING TO ADD

I have now told the story of the revenge of art, but I must go back to fill a gap in the plot. The hunt for asymptotic freedom involves looking at all possible theories—a task, it appears, that one would have to devote a lifetime to. Fortunately, Nature is once again kind to us. It turns out that there are not that many theories to be looked at. Let me explain.

In the chapter on quantum physics, we learned that to study a given theory, we have to add up the amplitudes of all possible histories. Now, that is an awful lot of adding. A particle, or a field, as the case may be, is faced with an infinity of possible histories. How in the world can one do the addition? Those of us who aspire to be fundamental physicists must spend a considerable amount of time learning how to add an infinite number of amplitudes together.

Let me give the reader a flavor of what is involved by adding an infinite number of ordinary numbers together. Suppose we are told to add $1 + 2 + 3 + 4 + 5 + \cdots$. The dots indicate that we are to keep on going indefinitely. Obviously, the sum makes no sense; it gets larger and larger, approaching infinity. Suppose instead, we are told to add $1 + \frac{1}{2} + \frac{1}{3} + \frac{1}{4} + \frac{1}{5} + \cdots$. In this case, as we keep going, the number we add on gets smaller and smaller. Nevertheless, it turns out that the sum still approaches infinity. Even though each number we add on gets smaller and smaller, the accumulated effect is huge. But now, suppose we are told to determine $1 - \frac{1}{2} + \frac{1}{3} - \frac{1}{4} + \frac{1}{5} - \cdots$. Here we may have a chance. The alternating positive and negative numbers cancel each other to some extent, and the sum grows very slowly. Just after adding 1/3,943, for example, to the sum, we subtract 1/3,944 the very next step. The net effect is to have added the tiny number 1/15,551,192. The sum, in this case, barely increases after a while. In fact, it eventually approaches a well-defined number, equal to $0.693 \cdots$. We reach a sensible answer.

The moral of the story is, in order to sum an infinite number of amplitudes, the structure of the theory must be such that the amplitudes tend to cancel. The cancellation, furthermore, is quite delicate. In the example above, if we go around switching some of the minus signs to plus, even if we switch only one sign in every hundred, say, the total sum will no longer be sensible but would once again approach infinity. When we consider a quantum field theory, the infinitude of histories is such that the addition of all these amplitudes makes sense only if the action has very special properties.

THINK LIKE ME

This is absolutely fabulous. A priori, physicists can write down an infinite variety of actions. Imposition of symmetries cuts down the possibilities immensely, but in general there are still many possible theories. Now we learn that in most of these theories we cannot add the amplitudes up, just as we cannot determine $1 + 2 + 3 + 4 + 5 + \cdots$. A quantum field theory in which we can sensibly add up all the amplitudes is said to be "renormalizable." Most remarkably, there are only three or four possible forms the action could take, depending on how one counts. The hunt for asymptotic freedom was possible precisely because one has to look at only a handful of renormalizable field theories.

I find it mind-boggling that physics has managed to narrow down the possible form of the ultimate design. He presented us with a puzzle. "Guess My design," He says. But there are so many possibilities! Then He shows us symmetry, and He shows us quantum physics, with its funny rules about adding the amplitudes of all possible histories. "Look, if you think symmetry and quantum, there aren't that many possibilities!"

The question of whether one can sum the amplitudes of the infinite number of histories in a Yang–Mills theory occupied a number of theoretical physicists after the theory was proposed. Finally, in 1971, a brilliant young Dutch physicist, Gerard t'Hooft, showed that, indeed, one can: Yang–Mills theory is renormalizable. This discovery was of monumental importance, since it implied that Yang–Mills theory makes sense as a quantum field theory. In this chapter, we saw that this realization broke the impasse over the strong interaction. In the following chapter we will see that this discovery literally ushers in a new era in our understanding of Nature.

13

The Ultimate Design Problem

THE END OF SYMMETRY?

Physicists have been astonishingly successful in using symmetry to fathom His mind, almost as if they have found the language He prefers.

They began with parity and rotational invariances and arrived at exact non-abelian symmetry. They witnessed the triumph of exact symmetries over approximate symmetries. They traveled a long road, following the Burning Tiger. Now let them take stock.

Imagine a physicist in the early 1960s reflecting on the state of symmetries. He probably would be pessimistic about the future of approximate symmetries. Heisenberg's once-shocking notion now appears to be inherently self-limiting in its utility. The proton and the neutron, related by isospin, have almost the same mass. The eight baryons, related by the eightfold way, have more or less the same mass. If there is a symmetry more approximate than the eightfold way, the particles it relates would have masses so different that we might have great difficulty recognizing them as related. An architect whose client wants a round building designs a hexagonal one. He tells the client, "Look, it's almost round," but the client may have a hard time recognizing the approximate circular symmetry. In practical terms, a very crudely approximate symmetry, even if recognized, would not be useful.

A PROBLEM IN DESIGN

What about exact symmetries? Our stocktaking physicist could not imagine how they could possibly point the way to the ultimate design. Basically, the difficulty is that symmetry implies unity while the world exhibits diversity.

It is a problem in design. Consider a rug weaver. If he insists on exact circular symmetry, he will produce a very dull circular rug. The only possible design consists of a bunch of concentric circular bands. By loosening the symmetry, he admits more interesting designs.

Our world does not resemble the dull circular rug; it is interestingly diverse.

In music, also, the tension between unity and diversity is often palpable.

When I speak of diversity here, I am not speaking of the bewildering variety of macroscopic phenomena that we see around us. As discussed in Chapter 2, physicists had already managed to reduce macroscopic phenomenology down to manifestations of electromagnetic interaction. Here, I am speaking of the diversity of the world at the particle-interaction level. The four interactions —strong, electromagnetic, weak, gravitational—have enormously disparate coupling strengths, and they differ totally in their characteristic properties. A motley crew of leptons and quarks participates in one or more of these interactions. The particles are full of individual quirks. The quarks interact strongly to form hadrons, the leptons do not. The electron is two thousand times lighter than the proton; the neutrino is massless. The electron interacts electromagnetically; the neutrino does not. And so on. Indeed, it is precisely this diverse array of properties that makes possible the structure of the observed world.

SYMMETRY VERSUS DIVERSITY

The diversity exhibited by the fundamental interactions and particles would appear to indicate, in the late 1960s, that perfect symmetries have no place in Nature's design. Yet, the intellectual disciples of Einstein shuddered to think that the Good Lord would prefer approximately symmetrical actions to perfectly symmetrical ones. Apparently, they must be resigned to thinking that there is no unifying symmetry in the action of the world, pulling together the four fundamental interactions. The strong interaction is a hundred times stronger than the interaction next in strength, namely the electromagnetic. To hope for an approximate symmetry tying these two interactions together appeared as untenable as arguing that an ellipse in which one axis is one hundred times longer than the other is approximately circular.

For the Ultimate Designer, herein lies the rub. Symmetry is beauty, and beauty is desirable. But if the design is perfectly symmetrical, then there would be only one interaction. The fundamental particles would all be identical and hence indistinguishable from one another. Such a world is possible, but it would be very dull: There would be no atom, no star, no planet, no flower, and no physicist.

Our pessimistic physicist feels that neither approximate nor exact symmetries will tell us much more about the ultimate design. To him, the idea of symmetries appears to have exhausted itself. Perhaps the Burning Tiger can only guide us so far.

Hey, wait a minute, you might say. We just saw Yang and Mills going off in a spectacularly new direction. Instead of looking for ever more approximate symmetries, they insisted that Nature uses exact symmetries. Years later, a hidden exact Yang–Mills symmetry indeed was uncovered in the strong interaction. So, why was our physicist bearish on exact symmetries?

True, our physicist did not foresee the possibility of an exact symmetry hidden in the strong interaction, yet the reasoning that led to his pessimism was basically correct. The exact gauge symmetry in the strong interaction forces the gluons to be massless and indistinguishable from each other. It relates quarks with the same flavor and hence the same mass, but not differently flavored quarks. By its very exactness, an exact symmetry cannot relate particles with different properties. Thus, in the absence of a drastically new concept, exact symmetries cannot tie together the disparate interactions.

AN IMPOSSIBLE DEMAND

This dichotomy between symmetry and diversity strikes deep at our aesthetic sensibilities. Perfect symmetry evokes repose, austerity, even death. Geometry inspires awe, but not exuberance. Modern sculpture strikes me as a continuing struggle to reconcile the geometric with the organic. Thomas Mann expressed this dichotomy in his *Magic Mountain*. Hans Castorp, a character in the book, nearly perished in a snowstorm. The snowflakes appeared to him as:

myriads of enchanting little stars, in their hidden splendor . . . , there was not one like unto another. . . . Yet each in itself—this was the uncanny, the anti-organic, the life-denying character of them all—each of them was absolutely symmetrical, icily regular in form. They were too regular, as substance adapted to life never was to this degree—the living principle shuddered at this perfect precision, found it deathly, the very marrow of death —Hans Castorp felt he understood now the reason why the builders of antiquity purposely and secretly introduced minute variations from absolute symmetry in their columnar structures.

We, too, prefer the world to deviate from the icy perfection of absolute symmetry, so that widely disparate interactions can play against each other, producing a world of interesting diversity, a world of organic beauty. But we cannot bring ourselves to imagine Him preferring ellipses to circles and introducing minute variations on the sly.

The Ultimate Designer wants both unity and diversity, absolute perfection and boisterous dynamism, symmetry and lack of symmetry. He appears to be putting an impossible demand on Himself.

THE WISDOM OF THE WINE BOTTLE

I consider it a dazzling tribute to the human intellect that physicists may well have discovered how He solved this impossible design dilemma. I will now explain.

When we speak of symmetry, we are apt to think of the symmetry of geometrical figures, such as those used in design and in architecture. A geometrical figure either has a certain symmetry, or it doesn't. The solution to the Ultimate Designer's problem cannot be visualized by thinking of geometrical figures.

The solution can be found, instead, at the bottom of a wine bottle (after the contents have been consumed). The bottom of many resembles a hollow bowl, as illustrated in Figure 13.1 (*left*). Drop a small marble in the bottle; obviously, the marble will end up in the center. But some bottles have a kind of hump—known technically as a punt or kick—in the bottom, as illustrated in Figure 13.1 (*right*). Drop a marble in a punted bottle, and watch it eventually come to rest at some point on the rim of the punt.

Figure 13.1.   The wisdom of the wine bottle: *(left)* an unpunted bottle, *(right)* a punted bottle.

So what, you say, this is child's play, and rather silly at that. But we are actually learning something of fundamental importance about the nature of symmetry. If I take off the labels, both the punted and the unpunted bottles are symmetrical under rotation around the bottle axis. The unpunted bottle with the marble in it is still symmetrical under rotations around the bottle axis. In contrast, the marble in the punted bottle, by virtue of its position on the rim, picks out a direction. Rotational symmetry is "broken." The configuration of the marble and the bottle is not invariant under rotations around the bottle's axis. The marble in the bottle may be pointing east, or north by northwest, as the case may be. (This is the fundamental operating principle of the roulette wheel, of course.)

SPONTANEOUS SYMMETRY BREAKING

To understand the lesson, we must appreciate that when speaking of symmetries in physics, we think not of the symmetries of geometrical figures but of the symmetries of actions. Now comes the crucial point. Given an action, we still must determine the actual history followed by the system, be it a particle or the entire universe. If we are talking about quantum physics rather than classical, we have to determine the more likely histories. In either case, even though the symmetry transformations leave the action unchanged, they may or may not leave the specific history unchanged.

In the wine bottle analogy, nothing in the interaction between the marble and the punted bottle favors one direction over another. Yet, when the marble comes to rest, a specific direction is picked out. The interaction between the marble and the bottle corresponds to the action of the world, while the eventual position of the marble corresponds to the actual history followed by the world.

If the actual history followed is not invariant under symmetry transformations that leave the action invariant, as illustrated by the marble in the punted bottle, physicists say that the symmetry is "spontaneously broken." The notion of spontaneous symmetry breaking is in contrast to that of explicit symmetry breaking. If the action itself is only approximately symmetrical, physicists say that the symmetry is broken explicitly.

To understand explicit symmetry breaking, let's return to the wine bottle analogy. We ask the glassblower to slip up in his art so that the bottom of the bottle is not perfectly symmetrical; a small depression is located somewhere away from the center. Drop in a marble and it ends up in this depression. This situation corresponds to such approximate symmetries as isospin and the eightfold way. The action, just like the bottle, manifestly is not symmetrical. Explicit symmetry breaking is also described as breaking "by hand." One puts the symmetry breaking into the action.

The reader is probably asking, "What's so profound about all this?" Well, the wine bottle analogy is admittedly simplistic, and doesn't convey the full subtlety of the physics involved. How profound is the concept? Let me just say that the best minds in particle physics did not conceive of the notion of spontaneous symmetry breaking until the early 1960s. Earlier on, when confronted with a broken symmetry, physicists simply assumed that the action was not symmetrical.

Using spontaneous symmetry breaking, the Ultimate Designer can solve His problem: He can have a design which exhibits at once symmetry and the lack of symmetry! He can write down a perfectly symmetrical action, yet have the actual history be non-symmetrical.

SPONTANEITY IS NOT PUT IN BY HAND

Physics is full of situations that can be described as examples of spontaneous symmetry breaking. Let me give a rather well-known example that Heisenberg in particular studied. Look at a piece of magnetic ore. At the microscopic level, the atoms inside the ore are spinning perpetually. As explained in Chapter 3, the direction of spin of each atom defines an arrow. Each atom, therefore, may be thought of as carrying an arrow, a kind of miniature compass needle. In a magnetic material, the net effect of the electromagnetic interaction between the atoms is such that a force tends to make the arrows carried by two neighboring atoms point in the same direction, much as two bar magnets next to each other tend to point in the same direction. In fact, a magnet is nothing more (or less) than a piece of ore in which the zillions of arrows contained in it are all pointing in the same direction.

But the electromagnetic interaction is rotationally invariant, showing no preference for any particular direction. How is it, then, that in a magnetic substance a particular direction is picked out? The answer is obvious. Suppose we can arrange to have the zillions of arrows pointing every which way. Order is soon born of the chaos. Somewhere, a cluster of arrows will be pointing all in the same direction, more or less, and they would convince neighboring arrows to point in that direction also. Pretty soon, the zillions of arrows all end up pointing in the same direction. The rotational symmetry inherent to the underlying physics has been broken spontaneously.

A kind of spontaneous symmetry breaking may also be discerned in some social trends. For example, say that at one time people are indifferent to the kind of water they drink, but then, due to peer-group pressure, two individuals in social contact tend to drink the same kind of water. An individual's preference may be indicated by an arrow pointing toward that particular kind of drink. Peer pressure may be represented as an interaction tending to align the two arrows carried by two individuals. The simplest hypothesis would be that this interaction is rotational invariant: People do not have preferences of their own but tend to drink what their friends drink. The reader can fill in the rest of the story.

REPOSE AND EXCITEMENT

But these examples can only be suggestive. Fundamental physicists are not interested in actual objects, such as a marble in a wine bottle or atoms spinning in a magnet. They are interested in the fundamental action of the world and the configuration of the world.

Now, what in blazes do these physicists mean by that phrase "configuration of the world"? No, they are not talking about the actual makeup of the universe, the distribution and orientation of the galaxies, or anything directly observable. To explain this concept, I must first say a few more words about field theory.

As mentioned in the previous chapter, modern physics is described in terms of fields. Thanks to science fiction, "field" is now endowed with an aura of mysterious power, but the basic concept of field, invented by a mathematically uneducated bookbinder's apprentice, is quite simple. Recall that the electromagnetic field is characterized by the strength of the electric and magnetic forces that a charged particle would feel at any given point in spacetime. In other words, a field is characterized by a bunch of numbers at a specific time.

In studying fields, physicists follow a strategy of first describing the system in repose. For example, the electromagnetic field is in repose if it is zero everywhere; that is, if electric and magnetic forces are absent. In our marble in the bottle analogy, we first determine where the marble comes to rest. Then we can ask what happens if we give the marble a slight kick. By studying the way the marble rattles around, we learn about the bottom of the bottle. Similarly, physicists ask what happens if they give the electromagnetic field a "kick." The electromagnetic field "rattles" around; if the energy associated with the rattling is concentrated in small regions, they call the packets of energy photons. Physicists say that the electromagnetic field has been excited from repose; photons are referred to as excitations of the electromagnetic field.

By studying the nature of the excitations, physicists can learn about the action governing a given field theory. While it takes considerable dedication to master the full subtleties of modern field theory, this basic strategy is rather natural and perfectly easy to understand. It is the same type of strategy used by a child trying to find out about an unfamiliar object; she shakes and rattles the object and studies the nature of the excitations.

I can now state what physicists mean by the configuration of the world: It's a description of the world in repose.

In field theories studied prior to the 1960s, the fields are always zero in the state of repose, or the ground state, as it is called technically. This corresponds to the case of the marble in the unpunted bottle. In the state of repose, the marble is at zero distance from the center. The magnitude of the field corresponds to the location of the marble, as measured by its distance from the center of the bottle. Consider the symmetry transformations that leave the action invariant. Under these transformations, a field that is zero remains zero. In our analogy, rotations of the bottle around its axis leave the bottle invariant. If the marble is located at zero —that is, at the center—it remains at zero under these rotations.

But what if one of the fields is not zero in the state of repose? This corresponds to the case of the marble in the punted bottle; the marble is at a nonzero distance from the center when it comes to rest and thus picks out a preferred direction.

Call the field that is not zero in the state of repose the Higgs field for ease of writing. (Peter Higgs was among the physicists who studied spontaneous symmetry breaking.) By being nonzero, the Higgs field, just like the marble, picks out a direction.

Consider the symmetry transformations that leave the action invariant. These transformations, in general, will change the Higgs field. In the punted bottle, a rotation about the bottle axis leaves the bottle invariant, but changes the position of the marble. As previously explained, while the interaction between the marble and the punted bottle is rotational invariant, the configuration of the marble in repose is not. Similarly, those symmetry transformations that change the Higgs field are spontaneously broken.

The strategy is now clear. Physicists can start with a symmetrical action of the world, but a symmetrical action resembling a punted bottle rather than an unpunted one. After spontaneous symmetry breaking, the actual physical laws derived from the action would no longer be symmetrical. The action of the world, of course, must be designed so that certain cherished symmetries, such as Lorentz invariance, remain unbroken. As was noted before, one advantage spontaneous breaking has over explicit breaking is that the pattern of symmetry breaking is controlled by the action and not by the physicist.

THE STUDY OF NOTHING

Incidentally, the world in repose is the most restful place you can ever visit. It is empty of all particles; it's known to physicists as *the* vacuum. The particles that make up the stars and you and me are, of course, the excitations. The vacuum is the world with all the excitations removed.

In order to determine the pattern of spontaneous symmetry breaking, physicists have devoted considerable energy to the study of the vacuum, thus provoking the quip that fundamental physics has now been reduced to the study of nothing!

Having learned that it is possible to have an action shaped by perfect symmetries, yet have the manifestations of the action be totally nonsymmetrical, we will go on to discuss a dramatic development made possible by our understanding of spontaneous symmetry breaking. I am referring here to the epoch-making realization that, in fact, the weak and the electromagnetic interactions are related. To recount this story, I have to tell you a little more about the weak interaction.

A CORPULENT MARRIAGE BROKER

At first sight, it appears that the weak interaction cannot possibly be related to the electromagnetic interaction, it is so much weaker than the electromagnetic interaction. Recall the ghostly insouciance of the neutrino, and contrast that with the gregariousness of the photon, who hobnobs with anyone who is charged. The electromagnetic interaction is long-ranged, while the weak interaction is so extremely short-ranged that the weak interaction between two particles turns on only when the two particles are practically on top of each other. The range of the weak interaction is tiny even on the scale of nuclear distances, which is essentially determined by the range of the strong interaction.

Recall from Chapter 11 that the interaction between two

particles is understood as resulting from a mediator particle shuttling constantly between the two particles, the marriage broker trying to pull two interested parties together. The range of the interaction is determined by the mass of the mediator particle. Thus the nuclear interaction is short-ranged because the pion is massive. The extremely short range of the weak interaction may be explained if its mediator is much more massive than the pion. The mediator of the weak interaction is known as the "intermediate vector boson," denoted by the letter W. While Yukawa had already speculated in his classic paper on a mediator for the weak interaction, the W boson was only discovered a couple of years ago. The leaders of the experimental team responsible, Carlo Rubbia and Simon van der Meer, were awarded the Nobel prize in 1984. The mass of W is several hundred times that of the pion; as a marriage broker, W is too corpulent to get very far!

Although W was discovered only recently, particle physicists were able to deduce many of its properties soon after the structure of the weak interaction was established. In order for the weak interaction to have its observed properties, its "marriage broker" must behave in a prescribed way. Remarkably, W resembles, in some respects, the photon, the mediator of the electromagnetic interaction. For example, W and the photon spin at the same rate. But in other respects, W and the photon are strikingly different. The photon is massless, while W is one of the most massive particles known experimentally. When a particle emits a photon, parity is conserved, but since the weak interaction does not respect parity, when a particle emits a W, parity is, shockingly, not conserved. The situation is even more puzzling than Chadwick's encounter with Tweedledum and Tweedledee. Imagine meeting at a party two people who have identical facial features. But, one of them weighs next to nothing, the other is the most massive individual you have ever met. Are they related or not?

OF THE SAME STRENGTH

By the late 1950s, some physicists were already suggesting that the resemblance between W and the photon indicated that the weak and electromagnetic interactions are related somehow. The first obstacle to this interpretation was posed by the enormous disparity in strengths. In quantum physics, the strength of an interaction is measured by the probability amplitude that two particles separated by a certain specified distance would interact, as seen in Chapter 11. Since now we understand the interaction as being the result of a mediator going between the two particles, this probability amplitude is equal to the product of three probability amplitudes: the amplitude for one of the particles to emit the mediator, the amplitude for the mediator to get to the other particle, and the amplitude for the other particle to absorb the mediator. (In quantum physics, as in everyday life, the probability that a chain of events will occur is equal to the product of the individual probabilities for each event.) This fact suggests a way around the obstacle of disparate strengths. Perhaps the amplitude for a particle to emit a W is really no smaller than the amplitude for the particle to emit a photon, but W is so massive that the probability amplitude for it to pass from one particle to another is very small—it gets so "tired" that it's prone to just turn back. This would explain why the weak interaction is so much weaker.

This argument allows us to guess what the mass of W might be. Suppose that we guess that the amplitudes for a particle to emit a photon and to emit a W are the same. Then the relative strength between the weak and electromagnetic interactions is determined solely by the mass of W. Thus, we can simply figure out the mass that W must have in order to reproduce the observed ratio of strengths.

A SISTERHOOD OF GAUGE BOSONS

Several physicists, notably Julian Schwinger, Sidney Bludman, and Shelley Glashow, all went one step further: They guessed that the photon and the W boson are both gauge bosons of a Yang–Mills theory. To appreciate their boldness, we must note that in the late 1950s, the relevance of Yang–Mills theory was totally unclear.

Recall, in Yang–Mills theory one particle is transformed into another upon emitting or absorbing a gauge boson. This fact fits in naturally with the weak interaction. In radioactivity, the archetypical weak process, a neutron is observed to disintegrate into a proton, an electron, and a neutrino. Theorists picture the neutron emitting a W and transforming itself into a proton (or, at a more fundamental level, a down quark inside the neutron emitting

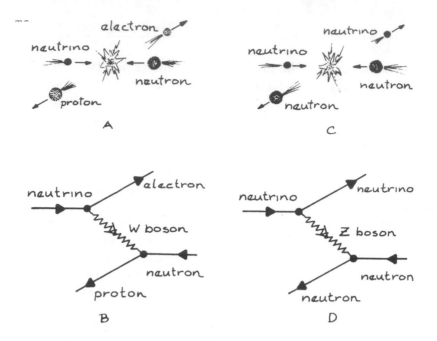

Figure 13.2. An artist's conception of a collision between a neutrino and a neutron. Two reactions could occur: (A) a charged current process, in which an electron and a proton come flying out, and (C) a neutral current process, in which a neutrino and a neutron come flying out. The charged current process was observed in 1961–62, the neutral current process in 1973. Instead of drawing pictures like those in A and C, physicists draw "Feynman diagrams," as indicated in B and D, to depict in more detail what is going on. In the charged current process (B), the neutrino emits a W boson and turns itself into an electron. The W then converts the neutron into a proton. In the neutral current process (D), the neutrino emits a Z boson and stays a neutrino. The Z is then absorbed by the neutron.

a W and transforming itself into an up quark). Or consider another typical weak process: A neutrino and a neutron collide, turning into an electron and a proton. See Figure 13.2. Again, we picture the neutrino emitting a W and transforming itself into an electron, and the neutron absorbing the W emitted by the neutrino and transforming itself into the proton. The study of weak interaction reduces to the study of what happens when a particle emits or absorbs a W.

At this point, we have to decide what group to use for the Yang–Mills theory. The choice of group determines the number of gauge bosons and their properties. Bludman tried the simplest group, our old friend $SU(2)$, but the resulting theory did not fit the observed pattern. Glashow persisted and went on to study the next simplest possibility, the group $SU(2) \times U(1)$. This group is basically just $SU(2)$, but with some additional transformations appended.

A LONG-LOST SIBLING

Glashow was able to fit the observed pattern of the electromagnetic and the weak interactions, but he also got more than he bargained for: The symmetry $SU(2) \times U(1)$ mandates an extra gauge boson, now called the Z boson.

When a neutrino emits or absorbs a Z boson, it remains a neutrino, according to the theory. (For that matter, an electron, a neutron, or any other particle remains itself when it emits a Z.) In this respect, the Z boson resembles the photon: When a particle emits or absorbs a photon, it does not change. But, in contrast to the emission or absorption of the photon, the emission or absorption of the Z violates parity.

The mediation of the Z produces a previously unknown interaction. For instance, when a neutrino and a neutron collide, this interaction could cause the neutrino and the neutron to simply scatter off each other, since the exchange of Z between the neutrino and the neutron does not cause either particle to be transformed. This process, now known as a "neutral current process," differs from the standard weak process in which the colliding neutrino and the neutron are changed into an electron and a proton. (See Figure 13.2.) The neutral current process clearly is even more difficult to detect than the standard weak one, since a ghostly neutrino comes out instead of an easily detected electron. Partly because of this difficulty and partly because of a widespread skepticism among the experimental community, the neutral current process was not actually detected until 1973.

I must emphasize that Einstein's dictum, symmetry dictates design, operates with full force. That there is an extra gauge boson leading to a previously unknown set of processes is forced on Glashow by the symmetry. Once he decided on a symmetry, he had no further say.

PRETTY BUT FORGOTTEN

The detection of the neutral current process vindicated dramatically the idea that a Yang–Mills theory can describe the electromagnetic and the weak interactions. But back in 1961, the situation must have looked rather discouraging. Experimenters had never seen a neutral current process. In addition, Glashow was faced with the apparently insurmountable difficulty that in a Yang–Mills theory, the gauge symmetry forces all the gauge bosons to be massless.

Not knowing what else to do, Glashow simply broke the gauge symmetry explicitly and put the masses of the W and the Z into the action by hand. By so doing, he lost predictive power, since he could put in essentially any masses he wanted. Since the W and the Z are in no sense approximately massless, he had to break the symmetry by an enormous amount. The resulting action is far from symmetrical. Remember the architect who tried to pass off a hexagonal building as an approximately circular one?

What's worse, breaking the symmetry by hand is a brutal process, upsetting completely the delicate cancellation that makes a theory renormalizable. Recall the discussion in the previous chapter on renormalizability and the problems of summing an infinite sequence of numbers. We can determine $1 - \frac{1}{2} + \frac{1}{3} - \frac{1}{4} + \frac{1}{5} - \cdots$, for instance. Breaking the symmetry by hand is akin to changing all the minus signs to plus, thus rendering the sum meaningless. In Glashow's work, there was no way one could sum over the infinite number of histories, so the theory made no sense.

Because of these difficulties, people stayed away from Glashow's work, and it was soon forgotten by all but a few who tried to keep the faith alive. In 1964, John Ward and Abdus Salam, whom we met in Chapter 11, attempted to resurrect the theory, but to no avail. Yeah, sure, Yang–Mills theory is kind of pretty, but what does Nature care about beauty anyway? The philistines are in ascendancy.

SPONTANEOUS BREAKING TO THE RESCUE

Meanwhile, the eminent Japanese-American physicist Yoichiro Nambu had introduced spontaneous symmetry breaking into particle physics. I mentioned earlier that Gell-Mann invented some symmetries of the strong interaction by using the "pheasant meat between two slices of veal" method. Gell-Mann's approach was so blatantly absurd that many physicists did not expect these symmetries to be relevant. And indeed, strong interaction phenomenology did not appear to be invariant under these symmetries. Later it turned out that these symmetries are present, in fact, in the action of the world; however, they are spontaneously broken.

Around 1964, with spontaneous symmetry breaking successfully applied to the strong interaction, various physicists— Philip Anderson, Gerald Guralnik, Carl Hagen, and Tom Kibble, François Englert and Richard Brout, and Peter Higgs—working in several independent groups, thought to look at what would happen if a gauge symmetry is spontaneously broken.

In gauge theory, you would recall, the gauge symmetries require the corresponding gauge bosons to be massless. Not surprisingly then, when a gauge symmetry is spontaneously broken, the corresponding gauge boson becomes massive. This today is called the Higgs phenomenon.

The fact that in a spontaneously broken gauge theory some gauge bosons become massive while others remain massless is just what the doctor ordered to cure Glashow's dying scheme! The W and the Z bosons could become massive while the photon remains massless. The whole idea just might work!

Surprisingly, at least in hindsight, Higgs and company did not apply their considerations to the weak and electromagnetic interactions. They treated their work as an amusing exercise in spontaneous symmetry breaking, and that was that. Glashow's scheme was not only dying, but also forgotten. The doctors who had the right medicine were nowhere near the patient. The psycho-social reasons for this curious turn of events are easy to understand if we place ourselves in the proper historical context. In the mid-1960s, the revenge of art was still a dream for the disciples of Einstein. The phenomenological approach dominated and gauge symmetries were far from the central concerns of the particle physics community.

Finally, in 1967, Abdus Salam and Steve Weinberg, working independently, came up with the brilliant idea of using the Higgs phenomenon to explain the differences between the weak and electromagnetic interactions.

At that time, Weinberg was using the notion of spontaneous symmetry breaking to study Gell-Mann's "veal-flavored pheasant meat" symmetries. He struggled to apply the Higgs phenomenon to these approximate symmetries of the strong interaction. As he recalled in his Nobel prize lecture, while driving to his office one day, he suddenly realized that he had been applying the right ideas to the wrong problem. (Incidentally, I generally avoid riding with theoretical physicists if at all possible, and when my wife and I go out, she always insists on driving. Absorbed in a problem, I once crashed my car on the road out of the Institute for Advanced Study in Princeton.) Once Weinberg realized the relevance of the Higgs phenomenon, he was able to work out quickly the unification of the electromagnetic and weak interactions.

Meanwhile, Salam had been pondering the symmetry properties of the weak interaction for years. I already mentioned that in 1964 he and John Ward had wrestled with the symmetry group $SU(2) \times U(1)$. The bout was apparently so bruising that Salam went on to work on other problems instead. As it happened, Tom Kibble, one of the physicists who discovered the Higgs phenomenon, was Salam's colleague at the Imperial College. As Salam recalled in his Nobel lecture, Kibble had tutored him about the Higgs phenomenon. In one of those unfathomable acts of creativity, Salam finally fused these diverse elements together in 1967.

The use of spontaneous symmetry breaking was crucial. Glashow had to put in the masses of the W and Z bosons by hand. This is analogous to the glassblower putting in a depression in the bottom of the wine bottle. With spontaneous symmetry breaking, on the other hand, the theory tells us about the masses of the W and Z bosons.

Interestingly, both Salam and Weinberg had been quite familiar with spontaneous symmetry breaking for some time. Weinberg had spent the academic year 1961–1962 at the Imperial College as Salam's guest. Together with the English physicist Jeffrey Goldstone, they had worked to elucidate spontaneous symmetry breaking. Curiously, years passed before Salam and Weinberg realized the relevance of spontaneous symmetry breaking to the unification of disparate interactions. The reason is clear, I think: As Weinberg has suggested, he and others, following Nambu, had focused on breaking the approximate symmetries of the strong interaction.

The work of Salam and Weinberg did not excite the theoretical community immediately. In fact, I remember that when I came across Weinberg's paper as a beginning graduate student, I was discouraged by one of my professors from reading it. Fundamental physics was dominated by the phenomenological approach, as I have repeatedly remarked. There was also another reason for the inattention of the theoretical community. No one actually knew how to sum the infinite number of histories in a Yang–Mills theory.

The Dutch physicist 'tHooft finally demonstrated how to do the sum in 1971, as was mentioned in the preceding chapter. The use of spontaneous symmetry breaking is again crucial. The delicate cancellations, which enable the sum to be done, would be destroyed were one to break the symmetry brutally by hand. I remember how excited my colleagues and I all got when news of 'tHooft's work reached us from Europe. Sidney Coleman proclaimed "'tHooft's work turned the Weinberg-Salam frog into an enchanted prince!" A theory of the electromagnetic and the weak interaction was finally at hand, now known simply as the standard theory.

A NEW EPOCH

The standard theory was a watershed development in the history of physics. It opened up a new epoch in our understanding of Nature. Physicists had reduced all physical phenomena to four fundamental interactions, so disparate that it appeared, at first sight, that no symmetry could possibly connect them. But Nature was only trying to fool us, hiding the elegant symmetries of the action. The photon, the W, and the Z indeed are related, as the gauge bosons of Yang and Mills, transforming into each other under the symmetry group. Like members of an identical triplet separated at birth, they retain only a vague remembrance of each other after spontaneous symmetry breaking. However, we can imagine physical processes involving energies much larger than the masses of the W and the Z, so that W and Z may be considered, effectively, as massless. In these processes, W and Z claim their kinship to the photon. Yes, we are your sisters, and we couple to particles just as strongly as you do! It is just that, at low energies, we are dragged down by our masses, so you think we are weak. Henceforth, the electromagnetic and weak interactions no longer exist as separate entities: They are unified into a single electroweak interaction.

The physicist I alluded to in the opening section of this chapter was too pessimistic about the power of symmetry. It was not the end of the road for symmetry, only the beginning.

14

Unity of Forces

WE THINK ALIKE

The unification of the electromagnetic and weak interactions marked the dawning of a new era in our understanding of Nature. The point is not that we have finally managed to understand radioactivity or the ghostly antics of the neutrino. Nor is the point that we now have a deeper understanding of electromagnetism. The point is, we are now emboldened to think that we can someday know His thoughts.

For the nearly forty years that approximate phenomenological symmetries ruled fundamental physics, believers in beauty and perfection plotted their comeback. The intellectual elements that went into electroweak unification were a long time in the making.

The quest began with Einstein's appreciation of symmetry, and his insistence on local transformations. The torch was carried by Weyl, who was moved by the deep spiritual truth uncovered by Noether. Heisenberg opened up a new world of internal geometries and symmetries. Yang and Mills built on this legacy. With the understanding of spontaneous symmetry breaking, these diverse elements finally came together in the "standard theory." That experiments have resoundingly vindicated this theory has had a stupendously liberating effect. Nature is telling us we are on the right track. She thinks about the same things that we, insignificant though we may be, think about.

A FATEFUL REUNION

Now that the electromagnetic and weak interactions have been unified into a single electroweak interaction, as described by a Yang–Mills theory, physicists wonder about the strong interaction. In Chapter 12, we saw how physicists discovered that the strong interaction is also described by a Yang–Mills theory. It was natural, therefore, to imagine that the electroweak and the strong interactions may be further unified with each other.

The photon, the W, and the Z have had a tearful reunion. They now look longingly at the eight gluons. Are you also our long-lost siblings? After all, we are all Yang–Mills gauge bosons. No, we cannot be born of the same womb, comes the reply. You are feeble, but we are strong! As messengers of the dark world inside hadrons, we live with power.

But wait! Asymptotic freedom speaks. Yes, indeed, you gluons are infrared slaves, confined by your own power, but as your energies increase, you yearn for freedom. At higher and higher energies, you gluons become weaker and weaker.

Yes, it is all coming together. In Chapter 12, when I recounted the search for asymptotic freedom, I mentioned that electromagnetism is not asymptotically free. In other words, as we look at the world with ever higher energies, the electromagnetic force becomes ever stronger, while the strong force becomes ever weaker. At some energy level, the electromagnetic force will become just as strong as the strong force. Unification is possible!

I, the photon, once thought, as I traversed the eternal loneliness of the universe, that I was all alone. I saw the W and the Z, in their massive obesity. How can I be related to them? But, in fact, they were not born overweight; their masses resulted only from spontaneous symmetry breaking. At high enough energies, the three of us are all massless. At energies even higher, we will become stronger, while you, the gluons, will become weaker. At some enormously high energy, it will be revealed that we are all siblings!

GRAND UNIFICATION

The dramatic proposal that the strong, the electromagnetic, and the weak interactions are unified at some energy scale is called "grand" unification, as distinguished from electroweak unification.

Around 1973 or so, the idea that perhaps we could go one step further and unify three of the four fundamental interactions was in the air. But, as we have learned, history is never cut-and-dried. There were still lingering doubts about electroweak unification. The skeptics criticized the entire theoretical framework as a house of cards not firmly grounded on actual observations. In theoretical physics, the prize goes to the bold, those who, while crossing a stream, do not wait to see if the next stepping-stone is firm before leaping. Often, they end up in the drink, but sometimes they get to the other bank before anybody else.

In 1973, Jogesh Pati and Abdus Salam, and, independently, in 1974, Howard Georgi and Shelley Glashow, came forth boldly with theories of grand unification. The two theories agree in general philosophy but differ in detail. The theory proposed by Georgi and Glashow is tighter and hence more predictive, and I will concentrate on it here.

A GREAT LEAP FORWARD

In a classic paper, Howard Georgi, Helen Quinn, and Steve Weinberg calculated the energy at which the fateful reunion of the strong and electroweak interactions would take place. Since we know how each coupling changes as the energy changes, it is a matter of elementary arithmetic to determine the energy at which they become equal. Recall, couplings move slowly—it takes a large change in energy to produce a small change in the couplings. In fact, they move so slowly with energy that they will become equal only at the fantastically large energy of 10^{15} times the nucleon mass. (The number 10^{15} is of course the mind-boggling 1,000,000,000,000,000.)

Physicists associate with each physical process a characteristic energy scale, defined as the energy carried by a typical particle participating in the process. (For example, the total energy involved in the collision of two Mack trucks may be quite awesome, but the typical energy carried by an individual nucleon in the truck is actually minuscule.) The energy 10^{15} times the nucleon mass is referred to as the grand unification energy scale. To appreciate how large this energy scale is, we may note that the characteristic energy released in a nuclear reaction is only about one hundredth the nucleon mass. Or, consider that at the world's largest accelerators, particles have been speeded up to energies of a few hundred times the nucleon mass—the largest energies ever produced by man.

Traditionally, physics has progressed by moving steadily

from one energy scale to the next. Here, by doing a simple calculation that would fit on a small piece of paper, theoretical physicists managed to leap forward to a dramatically new domain, where three of the four fundamental interactions are unified as one.

LONG-LOST SIBLINGS

The idea of grand unification is to bring together the photon, the W, the Z, and the eight gluons as the gauge bosons of a single Yang–Mills theory. The photon, the W, and the Z are the gauge bosons of a theory with the group $SU(2) \times U(1)$, while the gluons are the gauge bosons of a theory with the group $SU(3)$. Remember that $SU(3)$ is defined by transforming three objects into each other, and that $SU(2) \times U(1)$, or $SU(2)$, is defined by transforming two objects into each other. Now we are ready to perform one of the most important calculations in the history of physics: $3 + 2 = 5$. We conclude that we need a group that will transform five objects into each other. We want $SU(5)$.

Georgi and Glashow therefore proposed grand unification using a Yang–Mills theory with the symmetry group $SU(5)$. Once the group is specified, the number of gauge bosons is completely fixed by group theory. A head count shows that, in addition to the photon, the W, the Z, and the eight gluons, there are two additional gauge bosons, named simply the X and the Y. At the tearful reunion where the gluons finally recognized the photon, the W, and the Z as their long-lost siblings, two other individuals showed up. Later, I will explain the important roles possibly played by these two bosons in the evolution of the universe. Here, I would like to emphasize that the X and the Y are present whether we like it or not; group theory requires their presence. The situation is entirely analogous to the one encountered by Glashow when he found that the Z boson was necessary for electroweak unification.

LET THERE BE GRAND UNIFICATION

Let me summarize. Georgi and Glashow proposed that the Ultimate Designer started with a Yang–Mills theory based on $SU(5)$. At the grand unification scale, the symmetry is spontaneously broken into color $SU(3)$ and the $SU(2) \times U(1)$ of Glashow,

Salam, and Weinberg. In other words, the Yang–Mills theory breaks up into two Yang–Mills theories, one based on $SU(3)$, the other on $SU(2) \times U(1)$. At this stage, the X and Y bosons acquire enormous masses—on the order of the grand unification energy scale; that is, about 10^{15} times the nucleon mass—and bid farewell to their siblings, the gluons, the W, the Z, and the photon, which remain massless. As we come down in energy, we reach the electroweak energy scale, at an energy of a few hundred times the nucleon mass. The Yang–Mills theory, based on $SU(2) \times U(1)$, in its turn suffers spontaneous breakdown, whereupon W and Z become massive while the photon remains massless. Of all the sibling gauge bosons in $SU(5)$, only the photon and the eight gluons appear as massless excitations at low energies. The gluons are confined in infrared slavery, leaving only the photon to roam, shedding light on our world.

When He supposedly said, "Let there be light!" perhaps He actually said, "Let there be an $SU(5)$ Yang–Mills theory with all its gauge bosons, let the symmetry be broken down spontaneously, and let all but one of the remaining massless gauge bosons be sold into infrared slavery. That one last gauge boson is my favorite. Let him rush forth to illuminate all of my creations!" It doesn't sound as dramatic, but it is probably closer to the truth.

Figure 14.1.   A Blakian God bringing light into the world.

A SEAMLESS FIT

What about the other fundamental particles in the universe, the quarks and the leptons—how do they fit in? Under the $SU(5)$ transformations, the quarks and leptons are supposed to transform into each other, or, using math talk, the quarks and leptons are supposed to furnish representations of $SU(5)$.

What are the dimensions of some representations of $SU(5)$? Recall that the dimension of the representation is simply the number of entities belonging to that representation. The defining representation is five-dimensional, of course. In Chapter 9, we learned to construct larger representations by gluing representations together. Let us, then, glue two defining representations together.

Using the pictorial device from Chapter 9, we imagine gluing a circle and a square together. The circle and the square are to be colored with one of five possible colors. Since we have five color choices for the circle and five choices for the square, we have $5 \times 5 = 25$ combinations or entities. In the appendix to Chapter 9, I explained how these entities are to be divided into the so-called even and odd combinations. Let us look at the odd combinations; that is, combinations of the form Ⓡ Ⓨ − Ⓨ Ⓡ ($R =$ red, $Y =$ yellow, and so on). In order for the combination not to vanish, we must use different colors for the circle and the square. Let us count. There are five choices of color for the circle; for each of these choices, we have four choices of color for the square. It appears that we have $5 \times 4 = 20$ combinations, but since Ⓡ Ⓨ and Ⓨ Ⓡ appear in the same combination, we must divide by 2 to avoid counting twice. In other words, if we reverse red and yellow in the combination given above, we obtain Ⓨ Ⓡ − Ⓡ Ⓨ, which is not a new combination but merely minus the combination we already have: Ⓨ Ⓡ − Ⓡ Ⓨ $= -($Ⓡ Ⓨ $-$ Ⓨ Ⓡ$)$. So altogether we have $20/2 = 10$ odd combinations. We have a ten-dimensional representation.

If the reader feels a bit unsure about this counting, perhaps the following mathematically identical problem would help. At an intercollegiate tennis tournament, one team arrives with five players on its roster. For the doubles match, how many different pairs can the coach send in? He can pair Mr. Red with Mr. Yellow, and so on. Since he obviously cannot pair Mr. Red with himself, he may think that there are $5 \times 4 = 20$ pairs, but since pairing Mr.

Red with Mr. Yellow is the same as pairing Mr. Yellow with Mr. Red, the coach in fact has only 20/2 = 10 distinct pairs.

Enough mathematics—let us return to the quarks and leptons and count them. For reasons that will be explained later, we leave the strange quarks out. The leptons consist of the electron and the neutrino. We have the up quark and the down quark, but remember that each quark flavor comes in three colors, and so we actually have $2 \times 3 = 6$ quarks. Next, remember from Chapter 3 that the neutrino, the culprit for parity violation, always spins left-handed. In contrast, the other particles can spin in either direction. In quantum field theory, one associates a field with each spin direction. In other words, the electron is associated with two fields, while the neutrino is associated with only one. (Weyl was the one who figured out this arcane way of associating fields with spinning particles, and these fields are sometimes known as Weyl fields.) We finally are ready to count all the fields associated with quarks and leptons. Six quarks plus the electron, each with two fields, make $(6 + 1) \times 2 = 14$ Weyl fields. Add the neutrino field and we have altogether $14 + 1 = 15$ fields.

But, heavens, that is exactly equal to $5 + 10 = 15$!! The quarks and leptons would fit exactly into the five-dimensional and the ten-dimensional representations of $SU(5)$!

As a fundamental physicist, I imagine Him doing precisely this sort of calculation, simple but profound, and assuredly not the sort that covers pages and pages with messy formulas and equations.

The seamless fit of the quarks and leptons into $SU(5)$ convinces me and many other physicists that the Good Lord must have used $SU(5)$ in His design. The group $SU(5)$ may not be the whole story, but it is doubtlessly part of the story. The fit is even more seamless than our simple counting suggests. If one now examines how each quark and lepton would respond under the influence of each of the gauge bosons in the theory, one finds that they respond exactly as they should. For instance, by asking how each quark and lepton responds under the influence of the photon, one determines the electric charge of the quarks and leptons. One finds that the group theory of $SU(5)$ gives precisely the correct charge. Thus, the electron has one unit of negative charge, the neutrino no charge, and so on. As a rough analogy, we may think of the fit in a jigsaw puzzle: Not only do the pieces lock together, but the picture also comes out perfectly.

In this way, grand unification solves one of the most profound mysteries of physics: Why does the electron carry a charge opposite but exactly equal in magnitude to the charge carried by the proton? This fact plays an enormously important role in making the world the way it is: Atoms, and, by extension, macroscopic objects, can be electrically neutral. Before grand unification, this question of why the electric charges of the proton and the electron are exactly equal but opposite was regarded as unanswerable. Indeed, it was the kind of question that the vast majority of physicists, more interested in how the electron would behave in this or that phenomenon, would not even consider asking. (The quark picture does not answer it either, but merely reduces it to asking why quark charges are related to the electron charge.) In the Georgi–Glashow theory, the exact equality of the magnitudes of the charges carried by the electron and the proton emerges naturally from the group theory of $SU(5)$.

RENDEZVOUS À TROIS

Many physicists, myself included, are willing to believe in the Georgi–Glashow theory on aesthetic grounds alone. But physics ultimately is to be grounded in empirical verification. Remarkably, the very notion of grand unification can be tested experimentally.

Picture a path going up a mountain (see the illustration on the next page). Since coupling strength is a number that can only increase or decrease, we can picture a moving coupling strength as a kind of hiker moving along the path. The strength of the coupling corresponds to the elevation of the hiker. The "strong" hiker starts out high and proceeds to hike down. The "electromagnetic" and the "weak" start out close to the base of the mountain and proceed to hike up. In physics, the coupling strengths change as we increase the energy scale with which we view the world. In our analogy, the hikers move as time elapses.

To grand unify the world, we require the coupling strengths to meet at one energy. If there are only two hikers, one going up, the other coming down, then of course they will meet at some point. But if there are three hikers, two going up, the third coming down, then generally they would not all arrive at one place at the same time.

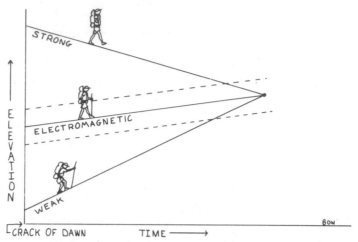

Figure 14.2. At the crack of dawn, a hiker named "strong" starts coming down a mountain while two hikers, named "electromagnetic" and "weak," start climbing up. "Weak" starts out lower than "electromagnetic" and has to move faster to keep up. I have plotted the elevations of three hikers as time passes. Given the starting positions of two of the hikers, the requirement that the three hikers arrive at the same point at the same time clearly fixes the starting position of the third. For instance, if "electromagnetic" starts out too high *(the upper dotted line),* she will run into "strong" before "weak" catches up. If she starts out too low *(the lower dotted line),* "weak" will pass her before he runs into "strong."

Suppose we know the rate at which each hiker moves as well as the starting positions of two of the hikers. Then the requirement that the three hikers all arrive at the same point at the same time clearly fixes the starting position of the third. Unless the third hiker starts at the exactly correct position, he will miss the rendezvous.

At this point, we realize that Nature is again kind to us. We have three interactions to be grand unified. Given the starting values—in other words, the values at low energies—of two of the three coupling strengths, we can predict the starting value of the third. The requirement that the world is grand unified, therefore, fixes the strength of the weak interaction relative to the strong and electromagnetic interactions. In practice, one uses this argument to predict the strength of the neutral current process. The experimental measurement agrees well with this prediction.

I mentioned in Chapter 2 and 10 that, in some instances, the functioning of the universe depends on a delicate balance between the disparate strengths of competing interactions. For a long time, physicists were perplexed by the presence of a hierarchy of interactions, so they were delighted to see grand unification explaining naturally the hierarchy between three of the four fundamental interactions.

THE COSMIC BOOK OF CHANGES

To seal the case for grand unification, we would have to see the X and the Y bosons. Considering that our most powerful accelerators can deliver an energy only a few hundred times the mass of the nucleon, we cannot hope to actually produce the X and Y bosons, which have masses on the order of 10^{15} times the mass of the nucleon. Instead, we can try to detect their effects. What do these bosons do?

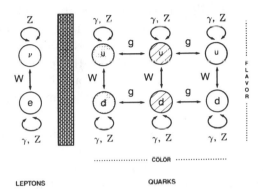

Figure 14.3. Late twentieth-century alchemy: A chart showing the transmutation of the constituents of matter by gauge bosons. The circles marked u, d, v (Greek *nu*), and e denote up quark, down quark, neutrino, and electron, respectively. Each quark comes in three different colors, suggested pictorially by the use of shading. The effects of the gauge bosons are represented by two-ended arrows labeled by the symbol of the corresponding gauge boson. As depicted, the W changes the up and the down into each other, but not its color. It also changes the neutrino and the electron into each other. In contrast, the gluons, g, change a quark into another quark with a different color but the same flavor. (For the sake of simplicity, we do not show all the gluons. Also, we show only two flavors, up and down; in the terminology of Chapter 15, we show only fermions belonging to the first family.) Thus, in this chart, the mediator of the weak interaction operates vertically, in the flavor "direction," while the mediator of the strong interaction operates horizontally, in the color "direction." When a particle emits or absorbs a photon, it remains the same particle, as indicated by the curved arrows which begin and end on the same particle. The photon is denoted traditionally by the Greek letter gamma, γ, as in gamma rays. Finally, the Z, like the photon, transforms a particle into itself. Notice that the neutrino is the only particle which does not interact with the photon. The X and Y bosons (not shown), postulated in grand unified theory, connect the worlds of quarks and leptons, currently separated as indicated by the brick wall. I imagine a medieval alchemist drawing a similar chart with one circle representing earth, say, another representing gold, and the arrow joining them labeled toad's blood. The difference is of course that our chart is based on facts.

To answer this question, let's first review what their siblings, the other gauge bosons, do. The gluons transform a quark into another quark with the same flavor but a different color; in other words, when a quark emits or absorbs a gluon, in its transformation it retains flavor but changes color. The gluons leave the leptons alone. The W boson, on the other hand, transforms a quark into another quark with the same color but a different flavor. It also transforms a lepton into a different lepton. For instance, the W boson transforms the electron into a neutrino. The photon transforms an electrically charged particle into itself. That is, when a charged particle, such as the electron, emits or absorbs a photon, it remains an electron. The photon leaves electrically neutral particles alone. Finally, the Z boson, like the photon, transforms a particle into itself, but, unlike the photon, the Z boson does not limit its interaction to charged particles. Confusing, isn't it? Perhaps Figure 14.3 helped.

According to modern physics, the ultimate reality in the physical world involves change and transformation. Here, a red up quark is changed into a blue up quark by a gluon, there, a blue up quark is changed into a blue down by a W. The W roams around and sees an electron. Poof, it turns the electron into a neutrino. It is a magician's world run wild.

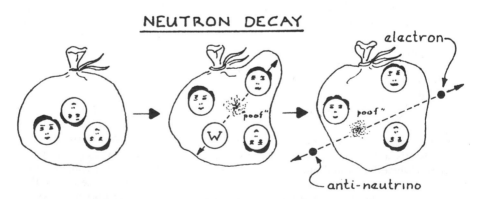

NEUTRON DECAY

Figure 14.4. Neutron decay and proton decay: The neutron consists of two down quarks, represented by the upside-down faces, and an up quark, represented by a rightside-up face, confined within a bag. Suddenly, one of the down quarks emits a W boson and turns itself into an up quark. The W then disintegrates into an electron and an antineutrino which, being leptons, escape from the bag. The two up quarks and the down quark left behind constitute a proton.

But in change there is permanence. Quarks are always changed into quarks, leptons are always changed into leptons. As I mentioned before, the transformations of quarks are manifested as the transformations of hadrons. As an example, consider the neutron, consisting of an up quark and two down quarks glued together by the gluons, and the proton, consisting of two up quarks and a down quark glued together. One of the down quarks inside the neutron may emit a *W* boson and transform itself into an up quark. As a result, we now have two up quarks and a down quark; in other words, a proton. The *W* boson, in turn, transforms an electron into a neutrino. What we actually observe is a neutron "decaying" into a proton, an electron, and a neutrino. (See Figure 14.4.) A neutron, if left alone, will actually do this in about ten minutes, on the average. The neutron, being more massive than the proton, can decay into the proton with energy to spare; that energy being imparted to the leptons. In contrast, the proton, being lighter, cannot decay into the neutron. We are led to an interesting question: If quarks can only be transformed into quarks, and leptons can only be transformed into leptons, something must be conserved. But what is being conserved?

Consider a magician whose art is limited to transforming one animal into another animal, one vegetable into another. A

PROTON DECAY

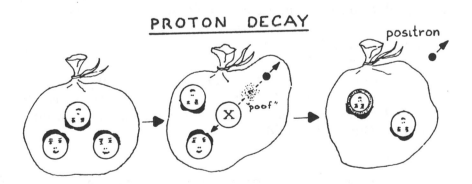

The proton decays when one of its up quarks suddenly disintegrates into an *X* boson and a positron. The other up quark absorbs the *X* boson and turns itself into an anti-down quark, represented by the upside-down face with a shaded rim. The positron escapes, leaving behind a bag containing a down quark and an anti-down quark, which we recognize as a neutral pion.

rabbit and an apple are on the stage. The magician, whose stage name is W. Boson, waves his cape, and, whoosh, the rabbit and the apple are transformed into a fox and some sour grapes. The audience bursts into applause. Whoosh, the fox and the grapes are gone, replaced by a mouse and a watermelon. But no matter how fantastic the transformations, there will always be one animal and one fruit onstage.

So, too, in the world of fundamental particles, the W boson is limited in his art. As a result, the proton is absolutely stable. The three quarks contained inside the proton cannot just vanish into thin air. The quarks inside the proton can only change into some other variety of quarks, but there will always be three quarks. Since the proton is the lightest hadron composed of three quarks, there is not a hadron for the proton to decay into. Protons are forever.

This is obviously good news. While everything around us is disintegrating and running down, the proton is a rock of solidity, guaranteeing the stability of the world.

What is conserved is the total number of nucleons, namely the protons plus the neutrons. The neutron could decay into a proton, but the number of nucleons cannot change. Actually, as we learned in Chapter 10, the hyperons, namely the cousins of the proton and the neutron via the eightfold way, also decay into the nucleons. Thus, strictly speaking, we must include the hyperons and speak of baryon number conservation. (Recall, baryon, as in baritone, is just the generic term used to refer to the proton, the neutron, and the hyperons. Each baryon is made out of three quarks.) Imagine putting twenty-one hyperons, four neutrons, and six protons in a box. Some time later, when we look in, we may find, say, ten hyperons, eleven neutrons, and ten protons. But no matter what happens, the baryon number remains unchanged at thirty-one. Quarks cannot disappear.

PROTONS, AND DIAMONDS, ARE NOT FOREVER

Into this reassuring picture crash the X and Y bosons. Before grand unification, quarks and leptons are kept apart; they belong to different representations. But in grand unification, the twelve quarks and three leptons are thrown into the five- and ten-dimensional representations of $SU(5)$. Twelve cannibals and three missionaries are to go into two boats, one capable of carrying five passengers, the other, ten passengers. Unavoidably, some quarks and some leptons are going to be thrown into the same representation. As a result, in a grand unified theory quarks can be transformed into leptons, and vice versa. The gauge bosons responsible are precisely the X and Y bosons. Quarks can disappear into leptons by interacting with X and Y bosons.

Onto the stage struts a new magician, Mr. X. Y. Bosons. Applause and, whoosh, the rabbit is transformed into an orange. Inside the proton, an up quark emits an X boson and changes into a positron. (The positron is the antielectron.) The X boson wanders over to the other up quark and, whoosh, changes it into an anti-down quark. What do we actually see? We start with two up quarks and a down quark, and we end up with a down quark, an anti-down quark, and a positron. Since quarks are enslaved, the down quark and the anti-down quark cannot emerge separately, but combine to form a pion. Thus, the proton disintegrates into a pion and a positron. (See Figure 14.4.)

THE FINAL DISASTER

Scientists are always telling us about impending disasters— the sun is going to explode in a supernova, engulfing poor planet Earth; our galaxy is going to collide with another galaxy; and so on. Those scenarios are bad enough, but here is real disaster, a disaster to outclass all other disasters, in the face of which a mere trifle such as the explosion of some star near the edge of some galaxy called the Milky Way pales into insignificance: Every proton in the universe is going to, poof, disintegrate. Stars will go out, our bodies will decay. Everything, for that matter, will decay into a cloud of pions and positrons. Matter will be no more.

But, do rest assured. This ultimate disaster is not going to happen for a while. Remember, the weak interaction is weak merely because the W boson is so massive. Now, if a corpulent 500-pounder can barely get around, imagine how a person weighing in at 5×10^{15} pounds feels. The effects of the X and Y bosons are going to be zillion times weaker than any weak interaction effects. A calculation using the $SU(5)$ theory shows that the lifetime of the proton is about 10^{30} years.

The mind reels before a timescale like this, a timescale that makes an eon look like a wink of the eye. I cannot really grasp how long ago the dinosaurs roamed the earth, so how can I understand the lifetime of the proton? I can't. I can only throw some numbers at you. The evidence is quite good that the universe is about 10^{10} years old. There are about 3×10^7 seconds in a year. Thus, the universe is about 3×10^{17} seconds old. Think of all the seconds that have ticked by since the universe began. Now, imagine expanding each second into the age of the universe. The total time elapsed would still be "only" $3 \times 10^{17} \times 10^{10} = 3 \times 10^{27}$ years, three hundred times shorter than the predicted lifetime of the proton. The proton lasts a long time.

HOW CLEVER HE IS

We see how clever the Ultimate Designer is. He wants grand unification, but He arranges for the coupling constants to move very slowly so that they meet at an extremely high energy. While the neutron decays in about ten minutes, the proton lives on beyond the eons. It would not be fun to create a universe that lasts only ten minutes! Talk about an unethical psychosociological experiment in which siblings are separated and raised under different circumstances. While the photon blithely dances, the hopelessly overweight X and Y bosons groan. Sorry, says the Boss, I have to keep you two overweight so that My universe can last for a while!

INTO THE SALT MINE

It would appear totally out of the question to verify experimentally that the proton actually decays. But thanks to the quantum law of probability, the experiment can in fact be done. In the quantum world, the statement that the proton has a lifetime of 10^{30} years indicates that, on the average, a proton will exist for 10^{30} years before disintegrating. Quantum physicists and insurance executives use the word "lifetime" in the same sense. A given proton has a small, but nonzero, probability of decaying the very next instant. It follows that if we could gather together enough protons, we may see one in the act of decaying. Indeed, if we watch an assembly of 10^{30} protons for an entire year, we should see one of them decay. Macroscopic matter, fortunately, contains an enormous number of protons.

When a proton in macroscopic matter decays, the pion and the positron, into which the proton decays, crash into the surrounding atoms, thus producing a telltale burst of light. What is the least expensive material that is transparent to light? Water, of course. In principle, then, an experimentalist need only fill a large enough tank with water and watch over it with sophisticated electronic cameras. In practice, the operation is considerably more difficult. The surface of the earth is being bombarded continually by cosmic rays, streams of particles that have been accelerated to high energies by magnetic fields in the galaxies. Cosmic ray particles crashing into water produce a background of light that, while too dim for human eyes to see, would totally overwhelm the minuscule amount of light associated with decaying protons. The only solution is to move the tank into a mine deep underground. Most of the cosmic ray particles will not be able to get through the layers of earth.

At this point, mining executives began to receive letters from various experimental physicists, outlining proposals for testing the ultimate stability of the universe. At the moment, experiments to detect proton decay have been set up around the world. One of the largest, involving several thousand tons of water (containing more than 10^{33} protons), is in a salt mine near Cleveland operated by Morton-Thiokol, Inc., a well-known salt producer in the United States. Other experiments are being performed in the Kolar Gold Field in India, the Caucasus Mountains in the Soviet Union, the Mont Blanc tunnel between Italy and France, an iron mine in Minnesota, a silver mine in Utah, and a gold mine in South Dakota.

Several years ago, when I found myself in South Dakota, I went to visit the experiment in the Homestake Gold Mine, which is known for the depth of its shafts. For a normally sedentary theorist, it was quite an experience. A safety officer from the mining company instructed me on various niceties not covered by Miss Manners, such as the correct way of walking in a mine. (One shuffles one's feet, so as not to trip in the dark.) I learned that in a mine, with its labyrinths of shafts and passages, the sense of smell provides the most efficient way of communication in an emergency. Bottles of a chemical with an exquisitely offensive smell are placed all around the mine. In an emergency, a bottle is smashed open and the warning smell is carried rapidly throughout the mine by the ventilation system, driven by gigantic fan mounted on the entrance of the mine. In the unpleasant part of the safety course, I had to take several practice whiffs of the warning signal. The elevator ride down is also memorable: The sensation is that of riding a New York City subway train traveling vertically in total darkness. Soon I found myself more than a mile underground, in a wet, windy darkness illuminated only by the miner's lamp on my helmet. The actual experimental area, however, was quite civilized, even equipped with such amenities as a refrigerator full of suitable refreshments for visiting theorists.

My experimental colleagues used mountaineering equipment to maneuver themselves about the experiment and diving equipment to get into the water. By not seeing any proton decay, physicists are able to set a lower limit to the proton's lifetime. Thus, if I watch 10^{30} protons for an hour and none of them decays, I can conclude that the proton's lifetime must be longer than 10^{30} hours. During my visit, most regrettably, I did not see a single proton decay. Remarking on this, my experimental friends joked that my visit represents a rare occurrence in the history of physics, when a theorist is able to advance human knowledge literally by doing nothing besides consuming refreshments.

Experimenters have watched a lot more protons for a lot longer time than I have. So far, they have not seen any proton decaying. Actually, bursts of light have been seen in the experiments, but they appear to be caused by the neutrinos contained in cosmic rays interacting with the nucleons in the water. A mile of rock would stop anybody, but not the ghostly neutrinos.

The lower limit on the proton lifetime is now in the range of 10^{31} to 10^{32} years. What are the implications for grand unification? The original version of $SU(5)$ grand unification predicted a proton lifetime of 10^{30} years and, thus, has been ruled out. But theorists have constructed several other versions in which the proton lives longer. Also, since the proton interacts strongly, an actual calculation of its lifetime has to contend with difficult details of the strong interaction.

The basic idea of grand unification is so overwhelmingly attractive that, at the moment, many physicists continue to believe in grand unification, even while conceding that the simplest realization of the idea may not be correct in detail.

THE DEATH AND BIRTH OF MATTER

If the proton is capable of dying, then it also must be capable of being born. If the proton can decay into a positron and a pion, then it follows that we can reverse the process and make a proton out of a positron and a pion. This simple remark opens up a dramatic new chapter in cosmology.

A FEW FACTS ABOUT THE UNIVERSE

There are two striking facts about the universe we live in: (1) The universe is not empty of matter; and (2) the universe is almost empty of matter. It is the task of fundamental physics to understand these facts.

Our image of the universe is one vast emptiness dotted here and there by a few galaxies. The philosopher Pascal was scared: *"Le silence éternel de ces espaces infinis m'effraie."* How do we measure quantitatively this frightening and almost inconceivable emptiness? How empty is the universe?

Matter is made of nucleons, but a mere count of their number in the universe does not define the paucity of matter in the universe; we have to compare the number of nucleons to some other number. It is natural to use the number of photons as a reference. It is now known that there are 10 billion (that is, 10^{10}) photons for every nucleon. In other words, matter is a one part in 10 billion contamination in an otherwise pristine universe. To fundamental physicists, a universe devoid of matter appears pure and elegant. I like to think of matter as the dirt in the universe.

GOD DOES NOT THROW DIRT AROUND

Before grand unification, physicists believed in absolute baryon conservation. The number of baryons in the universe— that is, the number of protons, plus the number of neutrons, plus the number of hyperons—cannot change.

Viewed in this light, it is doubly remarkable that the universe, while almost empty of matter, is actually not empty of matter. Suppose, for ease of talking, that there are exactly 537 baryons in the universe. Absolute baryon conservation would imply that the universe has always contained, and will always contain, 537 baryons, not one more, not one less. In that case, the issue of why the universe contains the actually observed amount of matter is not addressable by physics, but, instead, properly belongs to the realm of theological speculation. Whoever started the universe had to throw in 537 baryons.

In this picture, it is rather curious that Whoever threw in the dirt decided to throw in just a minuscule amount. Indeed, why would He want to throw in any dirt at all?

A MATTER-ANTIMATTER UNIVERSE

Faced with this conundrum, some physicists concocted a nifty solution: He did not throw in any dirt at all.

The idea is to exploit the proven existence of antimatter. Since the early 1950s, experimenters have observed routinely the production of pairs of particles and antiparticles. We count an antibaryon as having a baryon number minus one. Thus, the production of a proton and antiproton pair is perfectly consistent with absolute baryon number conservation.

The universe could start without any baryons; pairs of baryons and antibaryons could then be produced by the collisions of the particles that were present. No matter how complicated these production processes might be, absolute baryon conservation guarantees that there would always be an equal number of baryons and antibaryons. The matter and the antimatter got segregated, somehow, into different domains as the universe evolved. According to this view, we are wrong to think that the entire universe is constructed out of matter just because our immediate neighborhood is filled with matter. Perhaps the galaxy next to ours is made of antimatter.

The scenario that the universe is divided into matter and antimatter domains holds enormous appeal for science-fiction writers, but it does not hold up under scrutiny. Observationally, one might expect to see an occasional antiparticle in cosmic rays, interlopers from another domain. But they have not been seen. One might also expect that at the boundary between two domains, matter and antimatter would be annihilating furiously, emitting extremely energetic photons. Again, astronomers have not detected these telltale photons. Theoretically, the proponents of this scenario have never succeeded in finding a convincing mechanism that would segregate matter and antimatter. Thus, belief in baryon number conservation appears to preclude any understanding of the amount of matter in the universe.

A DILEMMA

The reason that physicists believed in baryon number conservation before grand unification is quite clear. The mere existence of matter implies that the proton's lifetime is larger than the age of the universe, and that seems like such a long, long time.

However, by the 1950s, some physicists had already felt somewhat uneasy about absolute baryon number conservation. Emmy Noether told us that a symmetry, either local or global, has to be responsible for baryon number conservation. In 1955 Lee and Yang pointed out that the symmetry responsible cannot possibly be local because the long-range effect of the massless gauge field required by the local symmetry would have been seen. Thus, for a physicist who subscribes to the aesthetic framework outlined in Chapter 12 and who views global symmetry with suspicion and distaste, exact baryon number conservation poses something of a philosophical dilemma. This dilemma was made all the more acute because the exact conservation of electric charge (which guarantees the ultimate stability of the electron) is indeed accompanied by a massless gauge field—namely the photon—as pointed out by Weyl.

And so, when it was shown that grand unification tosses out absolute baryon number conservation and denies the proton its immortality, some physicists felt a profound intellectual relief and satisfaction.

THE GENESIS OF MATTER

If the number of baryons is not absolutely conserved, then baryons could have been made in the early universe by physical processes. He did not have to throw in any dirt after all. The dirt generated itself. Grand unification opened up the possibility of understanding the genesis of matter.

Clearly, by itself, baryon nonconservation is not enough.

The fundamental laws of physics must distinguish, at some level, between matter and antimatter. If the laws of physics are totally impartial toward matter and antimatter, how could the universe have chosen to evolve into one containing matter rather than one containing antimatter?

For a long time, physicists believed that the laws of physics indeed did not distinguish between matter and antimatter. In Chapter 3, I explained that after the scandalous fall of parity, physicists continued to hope that Nature would still respect CP, the operation of reflecting particles into antiparticles, and vice versa, but that hope was dashed. In 1964, a minute violation of CP invariance was discovered in the decay of the K meson. For many years, it appeared as if CP violation does not affect any physical process other than K meson decay. But while we still do not have a good understanding of CP violation, we now have a clue to why He included a small amount of CP violation. He wants the universe to contain matter.

In this picture, the amount of matter contained in the universe depends on the extent to which CP invariance is violated. We now understand why the universe is almost empty of matter: CP violation is minuscule.

It appears, at first sight, that matter would not have had time to generate itself. The generation of a proton requires the intervention of the very same X and Y bosons responsible for the decay of a proton. One might think, therefore, that it would take some 10^{30} years for a proton to be born.

The resolution of this apparent paradox hinges critically on the notion of spontaneous symmetry breakdown. Imagine the X and Y bosons crying out to their siblings, Yes, we are hopelessly overweight and desperately weak compared to you, but at high energies, near the grand unification scale, our masses could be neglected and we too would be strong like you.

Shortly after the Big Bang, when the universe was extremely hot, particles were indeed zinging around with enormous amounts of energy. Even the X and Y bosons felt nimble. Processes in which baryons were generated occurred just as readily as electromagnetic processes. Matter was being born.

How do we know that the universe was extremely hot immediately after the Big Bang? We are all familiar with the fact that gases cool as they expand. For example, it gets colder as one climbs up a mountain. Our universe, similarly, cools as it expands. Knowing the temperature of the universe now, we can extrapolate backward to determine how hot the universe was at any given time in the past. In this way, it can be estimated quite easily that at about 10^{-35} seconds after the Big Bang, the energy of every particle in the universe was of the order of the grand unification energy.

Meanwhile, the universe continued to expand and cool. In a very short time, the energies of the X and Y bosons dropped below their enormous masses, and they became extremely feeble. Their moment in the sun, brief but glorious, was over. The baryons that were generated can live with quasi-immortality for the next 10^{30} years.

An expanding universe is absolutely crucial. In a static universe, the strength of the X and Y bosons would remain the same, and the birth and death of baryons would come to an equilibrium. Starting with no baryon, we would not be able to generate a net amount of baryons.

I am impressed tremendously by how cleverly He put it all together. Use the principle of local symmetry to produce grand unification with its inevitable violation of baryon conservation. Include a little bit of CP violation. Put in gravity, to make the universe expand. And, *voilà,* a universe that produces its own dirt, making possible stars, flowers, and human beings.

ORIGINS

An understanding of the genesis of matter is enormously satisfying. We humans have always wondered where everything comes from. In this century, that deep-seated quest for origins was reduced to the issue of how atoms were made. I stated in Chapter 2 that protons and neutrons were baked into helium nuclei a few minutes after the Big Bang. The more complicated nuclei were formed in stars and spewed out into space in stellar explosions. We, and everything else, are stardust, literally. Grand unification takes us one step further. We are, ultimately, the product of primeval forces mediated by the X and Y bosons.

In principle, the amount of matter in the universe can be calculated. Given a measurement of K meson decay, in fact, we should be able to predict, without ever looking outside the labora-

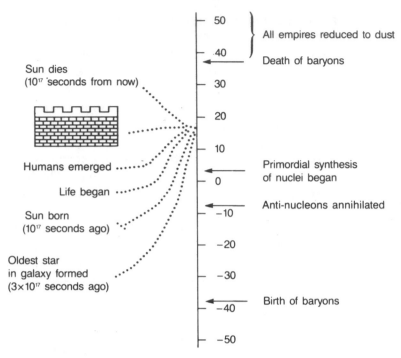

Figure 14.5. On the ruler of time, the number -30 indicates 10^{-30} seconds after the Big Bang; the number 30 indicates 10^{30} seconds after the Big Bang, and so on. In drawing this figure, I have assumed that the universe will go on expanding forever, as is indicated by current astronomical observations. The pictograph on the left of the ruler symbolizes human history.

tory, whether the universe is made out of matter or antimatter. Unfortunately, our understanding of *CP* violation is too rudimentary at present to allow us to carry out the calculation.

THE RULER OF TIME

We are born in the right time: We live in that epoch of the universe following the birth of the nucleons, but before their eventual death. I show in Figure 14.5 a physicist's history of the world. In order to incorporate the huge span of time involved, I plot time logarithmically. In other words, the notch marked 20 indicates a time 10^{20} seconds after the Big Bang. I list on the left side of the "time ruler" various events of particular interest to humans. A pictograph indicates the building of human empires.

STANDING ON A COOLED CINDER

The evolution of the world can be compared to a display of fireworks
that has just ended: some few red wisps, ashes, and smoke. Standing on
a cooled cinder, we see the slow fading of the suns, and we try to recall
the vanished brilliance of the origin of the world. . . .
—Lemaître

Our understanding of the genesis of matter opened up an
era in which grand unification breathes new excitement into cosmology. I explained that the universe gets hotter and hotter as one
extrapolates backward in time. As the universe gets hotter, the
typical energy of the particles in the universe increases. Thus, in
order to understand earlier and earlier epochs in the universe, we
have to master physics at ever-higher energies.

Before grand unification, cosmologists were limited to the
epoch that started, roughly, one millionth of a second after the Big
Bang. In one leap, physicists moved to the grand unification energy
scale. Cosmologists, correspondingly, are now able to track the
universe as far back as 10^{-40} seconds after the Big Bang.

Laymen are sometimes amazed that physicists claim to
know exactly what happened in the early universe. In fact, the
early universe, being a hot soup of particles zinging about, is considerably easier to describe than the universe at present, in which
the soup has splattered and congealed, so to speak. A vivid description of the early universe can be given once the physics at the
appropriate energy scale is established.

An account of the current research in early cosmology
would take us far beyond the scope of this book. I will limit myself
to mentioning one particularly exciting notion, that of an inflationary universe, as proposed by Alan Guth. Let us go back to the
punted wine bottle.

In the hot early universe, particles zing about with lots of
energies; this idea corresponds to the marble bouncing about in
the bottle. As the universe cools, the particles slow down. In the
bottle, the marble settles down to a state of repose and symmetry
is spontaneously broken. Now, suppose that there is a small
depression in the center of the punt. As the marble settles down,
it may be caught in the depression. The marble would possess
potential energy proportional to the height of the depression above
the bottom of the bottle. In a similar manner, the Higgs field may

also be trapped for a certain period of time, unable to reach its natural state of repose. The trapped Higgs field, like the marble, would possess potential energy.

The expansion of the universe is driven by the amount of energy contained within the universe. Crudely, we can think of the expanding universe as a balloon being blown up. The enormous energy carried by the trapped Higgs field causes the universe to expand so rapidly that the expansion can only be described as rampant inflation. It is estimated that during this inflationary epoch, the universe doubled its size every 10^{-38} seconds or so.

Here we return to some puzzling questions brought up earlier: Why is the universe so large? Why does it contain so many particles? Alan Guth pointed out that if the universe had once been in an inflationary epoch, these and other related questions could be answered. The universe got large through inflation. During inflation, the potential energy contained in the trapped Higgs field was rapidly converted into particles.

Although the actual implementation of the inflationary scenario has met with grave difficulties, the basic idea remains extremely exciting and appealing. It addresses and answers questions that several years ago would have been considered beyond the scope of physics.

The interface between particle physics and cosmology has emerged as one of the most exciting areas of research. At a conference on the early universe that I attended a few years ago, the participants took to giving their talks wearing T-shirts that proclaimed, "COSMOLOGY TAKES GUTS"—GUT being the acronym of grand unified theory. And those who do not believe in grand unification are dismissed by some as gutless!

NEW AND PERHAPS IMPROVED

In the years since grand unification was invented, theorists have constructed a number of other grand unified theories, seeking to improve on the Georgi–Glashow $SU(5)$ theory. For instance, many theorists consider the fact that quarks and leptons are assigned to five- and ten-dimensional representations, as unsatisfactory. They believe that in a truly unified theory, quarks and leptons belong together in one single representation.

Curiously enough, it turned out to be impossible to unify

the known quarks and leptons in a fifteen-dimensional representation. Instead, the search leads to the group $SO(10)$, which contains naturally the group $SU(5)$. But the group $SO(10)$ does not contain a fifteen-dimensional representation; rather, it contains a sixteen-dimensional representation, into which the known quarks and leptons fit naturally. Is group theory telling us that we have missed one extra field?

Recall that we counted fifteen fields associated with quarks and leptons, because the neutrino always spins left-handed. It is most intriguing that the extra field in the sixteen-dimensional representation of $SO(10)$ turns out to have precisely the right properties to be associated with a neutrino spinning right-handedly. Thus, the group theory naturally leads physicists to consider a grand unified theory based on $SO(10)$. At some energy scale, the $SO(10)$ symmetry breaks down spontaneously to $SU(5)$, whereupon the right-handed neutrino field acquires an enormous mass, thus explaining the fact that a right-handed neutrino has never been seen experimentally.

The theory also tells us that when the right-handed neutrino field acquires an enormous mass, the left-handed neutrino field is forced to acquire a minuscule mass. At present, a number of experimenters are actively trying to determine whether the left-handed neutrino, long believed to be exactly massless, actually has a minuscule mass.

Many theorists are inclined to believe in the $SO(10)$ theory, but, at the moment, it is far from being established experimentally. I mention the $SO(10)$ theory to give the reader a flavor of what grand unification research is like. The flavor is that of symmetry and group theory, of counting the number of fundamental fields, and of fitting them into the correct representations.

DESIGNER UNIVERSES

In Chapter 2, I told of theorists designing universes in their imagination. Now I have explained to the reader the rules of the game. Pick your favorite group: Write down the Yang–Mills theory with your group as its local symmetry group; assign quark fields, lepton fields, and Higgs fields to suitable representations; let the symmetry be broken spontaneously. Now, watch to see what the symmetry breaks down to. (In our wine bottle analogy, we watched for which direction the marble picked out in the punted bottle.) That, essentially, is all there is to it. Anyone can play. To win, one merely has to hit on the choice used by the Greatest Player of all time. The prize? Fame and glory, plus a trip to Stockholm.

Oops, I chose the wrong group and ended up without any massless gauge boson. Well, my universe would not have any light in it. No good. Try again. I choose another group, but now I end up with two massless gauge bosons. This universe would have two different kinds of photons. Well, another possible universe hits the wastebasket. Want to play?

LIVING IN THE DEBRIS

According to grand unification, we are living amid the debris of spontaneous symmetry breaking. True physics is at an energy scale of 10^{15} times the nucleon mass; the physics we observe represents only bits and pieces of this true physics. It is dizzying for me to think that the photon, whose behavior underlies the vast majority of macroscopic phenomena, is but one of many gauge bosons of the true physics.

To appreciate the role of spontaneous symmetry breaking, suppose, for the moment, that the Good Lord had broken the symmetry by hand. The architectural analogy would be to take an edifice built with a supremely intricate symmetry, and demolish it into rubble. Physicists may then be compared to intelligent ants crawling in the rubble, trying to reconstruct the original design. Physics would be forever doomed to be phenomenological. But, the Ultimate Designer appears to have broken the symmetry spontaneously instead—and that is of crucial importance in enabling us to glimpse the true physics even while we are limited to pitifully low energies.

15

The Rise of Hubris

TO SEE THE ENTIRE DESIGN

Throughout history, we physicists sought to understand one phenomenon after another; that is how physics progressed. Why does the apple fall down while the moon does not? What is this mysterious effect we call light? What is inside the atomic nucleus? But then, in one magnificent and unprecedented leap, fundamental physicists went from studying phenomena at energy scales a hundred times the nucleon mass or less to contemplating the physics at 10^{15} times the nucleon mass. The hubris of my generation of physicists knows no bounds. We have glimpsed how He designed the universe; now we imagine that we, too, can design universes.

The character of research in my field has changed drastically. I was in graduate school during the rule of phenomenology, when physicists grappled with such questions as how to calculate the collision of two protons. Many of these questions dealt with what Einstein called "this or that phenomenon," and they were never answered. Fundamental physicists simply ceased to care about these questions. They asked, and answered, much more profound ones: Why is the electron charge exactly equal and opposite to the proton charge? Why is the universe not empty of matter? Why is the universe so large?

Many physicists now feel that the big picture may be within our grasp, thanks to the guiding light of symmetry. After years of focusing on little patches of the oriental rug, we may finally be in a position to see the entire design.

A FLAVOR OF CURRENT RESEARCH

Even in their euphoria, fundamental physicists realize that they have not yet achieved a truly unified understanding of the physical world. To begin with, grand unification does not include gravity. Even leaving gravity aside, it is clear that the Georgi–Glashow theory is not the last word on grand unification. While certain long-standing fundamental questions have been answered, others remain as mysterious as ever.

In this chapter, I will try to give the reader a flavor of research in fundamental physics. I will focus first on one of the questions left unanswered by the Georgi–Glashow theory, then I will sketch some of the attempts to bring gravity into the fold.

THE IMPOSTOR

In 1935, working on the summit of Pikes Peak in Colorado, experimenters Carl Anderson and Seth Neddermeyer discovered a particle in the cosmic rays. At first, this particle was thought to be the meson discussed by Yukawa and now known as the pion. The new particle had a mass roughly equal to the value Yukawa had predicted for the meson, but, curiously, it did not behave at all like a mediator of the strong interaction. After a great deal of confusion, physicists realized that this particle, today known as the muon, was in fact not the pion; its mass just happened to be about the same as the pion's. Nature had tried to trick us.

Further studies revealed that the muon has exactly the same properties as the electron. The only difference between the two particles is that the muon is a couple of hundred times more massive than the electron. The muon is just a heavier version of the electron. Being more massive, the muon can decay into the electron via the weak interaction.

But what is the point of throwing the muon into the ultimate design? As far as we know, if the muon had been omitted, the universe would still function in the same way. The muon is redundant. It sits around for a while and then disintegrates into the electron. Aside from that, the electron can do anything the muon can do. In exasperation, the eminent experimentalist Isidor Rabi exclaimed, "Who ordered the muon?" Who, indeed? No one knows.

Curiously, the behavior of the muon under the weak interaction parallels that of the electron. The electron is transformed into the neutrino when acted upon by the W boson. Aping the electron, the muon is also transformed into a neutrino when acted upon by the W boson. A landmark experiment performed in the late 1950s established that the two neutrinos are not the same. To distinguish between them, physicists refer to one as the electron neutrino and the other as the muon neutrino.

In the 1960s, it slowly dawned on physicists that the strange quark is to the down quark as the muon is to the electron. The infamous strange quark has exactly the same properties as the down quark. Again, the only difference is that the strange quark is about twenty times more massive than the down quark. Nature is repeating Herself!

THE DISCOVERY OF CHARM

At this point, physicists made an obvious guess that there was also a heavy version of the up quark. But it was only a guess. In the late 1960s, Shelley Glashow, working with Greco-French physicist John Iliopoulos, and Italian physicist Luciano Maiani, showed that this additional quark, which they dubbed the "charm quark," is required in a Yang–Mills theory of the weak interaction. The structure of the gauge group is such that unless the charm quark is included, certain hadrons would decay in ways excluded by observations. That the theory predicts these unobserved decays had posed, for some years, a serious obstacle toward realizing a gauge theory of the weak interaction. The extra quark indeed worked like a charm in exorcising the unwanted decays.

The charm quark was discovered experimentally in 1974. I remember how excited everyone got. Together with the discovery of neutral current interaction, the charm quark discovery indicated that the theoretical schema, based on local symmetry and spontaneous symmetry breaking, was, indeed, correct. We think alike!

REDUNDANCY IN DESIGN

Goldilocks wandered into the bears' cottage and found that everything was triplicated. On the table were three bowls, identical in every way except size. Physicists are just as bewildered. They finally figured out how the universe is constructed: Matter is composed of the electron, the electron neutrino, the up quark, and the down quark; a bunch of gauge bosons and the graviton act on these quarks and leptons, transforming them into each other; out of this set up we get the entire splendor of the universe! A supremely elegant design, isn't it? But, just as physicists were about to swoon in admiration, the Ultimate Designer threw in an entire crew of particles that apparently play no essential role whatsoever in the healthy functioning of the universe. The electron is repeated in the muon, the electron neutrino in the muon neutrino, the up quark in the charmed quark, and the down quark in the strange quark. (To distinguish between these two crews of particles, physicists refer to them as the electron family and the muon family.)

The mystery deepened. Starting in the mid-1970s, experimenters discovered even more fundamental particles, and it became clear that there was a third family, consisting of something called the "tau," the tau neutrino, the top quark, and the bottom quark. The tau is an even heavier version of the electron, but behaves in every other way just like the electron and the muon. Similarly, the other particles in the tau family are corresponding replicas of the particles in the electron and muon families.

Nature is not only confusing physicists, She is taxing our ability to coin cute names! Top and bottom sound awfully close to up and down. (As I was writing *Fearful Symmetry,* experimenters claimed to have discovered the top quark. But by the time I got around to editing, the claim had been withdrawn.) Meanwhile, some theorists had speculated that perhaps there wasn't any top quark and constructed theories that became known, quite naturally, as topless theories. Many of us, however, thought that such nomenclature seriously compromised the dignity of the profession, and certain journals refused to put the term in print. The observed behavior of the bottom quark, fortunately, supports the existence of the top. For this and a variety of other reasons, most theorists believe in the top quark.

PAINTING FEET ON SNAKES

At the moment, physicists know of no good reason for including the muon and tau families in the design of the universe. Indeed, since the particles in these families decay rapidly into particles in the electron family, they are not even present in the universe, normally. The universe would function perfectly well if the muon and tau families were not even included.

Rabi's exasperated question is now updated to, "Why is He repeating Himself?" He appears to be ruining His own elegantly simple design with unnecessary embellishments. In China, the story is told of an artist greatly skilled in painting snakes. His work was much admired, but, nevertheless, he was not satisfied. The snakes he painted just did not look right to him. Finally, picking up his brush, he painted feet on the snakes. The Chinese expression "painting feet on snakes" is now used to describe the destruction of a design due to excessive embellishments.

Did Nature paint feet on snakes? Physicists do not think so. The prevalent faith is that in Xeroxing the matter content of the universe, She must have been motivated by a deep aesthetic imperative which we are yet unable to appreciate.

Figure 15.1. Painting feet on snakes: A contemporary cartoonist reinterprets an ancient Chinese moral tale, saying more or less that you should quit while you are ahead.

THE FAMILY PROBLEM

Physicists sometimes describe the electron, the muon, and the tau families as three generations in one big family. The puzzle as to why Nature should include three generations, when one would have sufficed, is known as the family problem. A few years ago, I was invited to give some lectures in Japan. When I spoke of the family problem, the audience burst out laughing. It turns out that the real-life family problem caused by three generations living together was, at that time, a hot topic in the Japanese news media!

Grand unification sheds no light on the family problem. Recall that when I counted fifteen quark and lepton fields in the last chapter, I omitted the strange quark. Now you understand why I did that: I wanted to count one generation at a time. Each generation contains fifteen quark and lepton fields, fitting nicely into a five-dimensional representation and a ten-dimensional representation of $SU(5)$. To accommodate three generations, Georgi and Glashow simply triplicated the representations appearing in their theory. But we have absolutely no understanding whatsoever as to why the representations are to be triplicated and why each successive generation is more massive.

The family problem is one of the deepest outstanding puzzles in physics today. At the moment, we can't even say for sure that there are just three generations; perhaps there are more. A number of physicists have tried to fix the number of families from first principles. In this endeavor, the Burning Tiger again leads the way.

MIRRORS

Many theorists assume that a group of symmetry transformations relates the different families. One may feel that in a truly grand unified theory, all known quarks and leptons should belong to one single representation which upon some kind of spontaneous symmetry breaking, decomposes into copies, perhaps three, perhaps more, of the five- and ten-dimensional representations of $SU(5)$. Intriguingly, this is possible only if additional particles that are mirror images of the known quarks and leptons are present. Group theory forces us to introduce a mirror electron, a mirror neutrino, and so forth. The mirror particles behave as if they are the mirror images of the known particles. For instance, the W boson transforms the electron into a left-spinning neutrino, but it transforms the mirror electron into one that spins right. Since experimenters have never observed mirror particles, mirror quarks and leptons, if they exist at all, must be more massive than the known quarks and leptons.

This brings us to the intriguing possibility that the ultimate design may be, in fact, parity invariant, and that the parity violation that shocked the physics world in the 1950s is but the result of a spontaneous symmetry breaking. Did He include mirror particles in His design and then proceed to break the mirror?

STANDING APART

In his definitive biography of Einstein, Abraham Pais wrote that he was struck by the apartness of the man. The gravitational interaction, in many ways Einstein's child, also stands conspicuously apart from the three other interactions.

Gravity appears singularly different from the other interactions, even leaving aside the vast disparity in strengths. Given that Yang–Mills gauge bosons mediate the other three interactions, one may guess that the graviton, the particle mediating gravity, is also a gauge boson. But it's not. The graviton behaves quite differently from a gauge boson; for instance, the graviton spins twice as fast as a photon. The graviton cannot be related directly to the mediators of the other three interactions.

During the reign of the phenomenological approach to particle physics, gravity often appeared as a forlorn stepchild, admired but neglected. Because gravity is so fabulously weak, its effects are completely negligible in the microscopic world. In those days, one could be a leading particle physicist without any understanding of gravity whatsoever. Even today, most physicists obtain their Ph.D.'s without bothering to take a course on Einstein's theory of gravity. A physicist studying the electronic properties of solids, for example, would not have to pay the slightest attention to the theory of gravity.

Even now, opinions differ on the role of gravity. Some physicists feel that we can fully understand the grand unified interaction only by relating it to gravity. Others prefer to focus on grand unification without worrying about gravity. In any case, one overwhelming thrust in current research consists of efforts to draw Einstein's child into play.

A MARRIAGE PROPOSAL RESISTED

As a classical theory, Einstein's gravity is beautifully complete, but just as he stubbornly refused to subscribe to quantum physics, it has steadfastly resisted marriage with the quantum. When the quantum principles are applied to Einstein's theory, the resulting theory of quantum gravity does not make any sense: Quantum gravity is not renormalizable. In other words, when physicists try to sum the infinite number of amplitudes associated with a gravitational process, they encounter a sum akin to $1 + 2 + 3 + 4 + \cdots$.

A spectrum of opinion surounds this issue. One extreme holds that Einstein's child is telling us that quantum physics must fail at some point. Others feel that the theory of gravity must be modified. Who is spurning whom?

Figure 15.2.   A marriage proposal refused: Einstein's theory of gravity spurns the quantum.

Physics started with gravity, but, ironically, it may also end with gravity. Of the four fundamental interactions, gravity is the least understood.

EINSTEIN'S QUEST

The classical world of Einstein admitted only electromagnetism and gravity, and Einstein was quite convinced that the two were related, particularly after Weyl had demonstrated that electromagnetism, like gravity, is based on a local symmetry. After his great work on gravity, Einstein devoted his scientific energy to a quixotic quest for the so-called unified field theory, a quest that some biographers view as tragic.

To his contemporaries, Einstein's quest appeared boneheaded and misguided. As Einstein labored, the world became quantum. The weak and the strong interactions were discovered, and phenomenology came to rule fundamental physics. It appeared absurd and terribly old-fashioned to insist on the unification of electromagnetism with gravity when the world contains two other interactions that appear to have nothing to do with local symmetries. Laughing at Einstein's futile labors, Pauli once quipped, "Let no man join together what God has put asunder."

But Einstein had the last laugh on Pauli. In some sense, grand unification realizes Einstein's impossible quest. Physicists have joined together what God has only appeared to put asunder. While it is true that unification of the other three interactions, leaving gravity out, was quite different from what Einstein had in mind, his vision of a unified design continues to inspire us today.

I began this book by saying that physicists are sustained by their faith that Nature, ultimately, is simple and comprehensible. The drive toward simplicity and unity has now arrived at a grand unified theory of the strong, the electromagnetic, and the weak interactions (see Figure 15.3). Only gravity remains outside the fold. Fundamental physicists are titillated by the thought that perhaps only one more step separates them from the ultimate design.

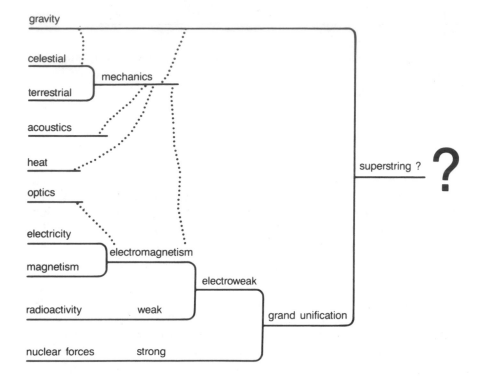

gravity

celestial

terrestrial

mechanics

acoustics

heat

optics

electricity

magnetism

electromagnetism

electroweak

radioactivity

weak

grand unification

nuclear forces

strong

superstring ?

?

DRIVE TOWARDS UNITY

Figure 15.3. The drive toward unity late in the twentieth century. (Compare with Figure 4.6.) Fundamental physicists are titillated by the thought that perhaps only one more step separates them from the Ultimate Design. Is superstring that final step? Opinions differ.

DIMENSIONS OF THE WORLD

It is ironic, perhaps, that the current drive to unify gravity with the other three interactions relies on an idea resurrected from the dustbins of history. In 1919, just four years after Einstein had proposed his theory of gravity, Polish mathematician and linguist Theodor Kaluza came forth with the absolutely nutty idea that spacetime is actually five-dimensional. This idea was developed by Swedish physicist Oskar Klein into what is known as the Kaluza–Klein theory.

Einstein had married space with time, and he described the physical world as having four dimensions: the three familiar dimensions of space and the dimension of time. But to Einstein, space remained three-dimensional, as any ordinary person would have

thought. Kaluza and Klein were saying something much more rad-
ical. In their scheme, space itself is four-dimensional. (It follows
that spacetime is five-dimensional.)

How could we possibly have missed one extra dimension in
space, having lived in it all our lives? Did Kaluza and Klein mean
to tell us that there is another direction in which we can move?

To understand the answer to these questions, consider a
creature constrained to live on the surface of a long tube. An
observer could see that the space "inhabited" by the creature, the
surface of the tube, is really two-dimensional. But suppose the
radius of the tube is much smaller than the smallest distance that
can be perceived by the creature. To the creature, space would
appear to be one-dimensional, since the creature can move only
along the tube. The creature would think that it was living in a two-
dimensional world: one dimension for time, another for space. In
other words, a very thin tube could be mistaken for a line. On
closer inspection, every "point" on the "line" actually turns out
to be a circle.

Kaluza and Klein supposed that every point in the familiar
three-dimensional space in which we move would, on closer inspection, also turn out to be a circle. If the radii of the circles are
much smaller than the smallest distance that we can measure, we
would see the circles as points, and we would be misled into thinking that we are living in a three-dimensional space instead of a
four-dimensional one.

The discussion thus far has to do only with geometry. Physics enters with the supposition of Kaluza and Klein that the world,
now imagined to be actually five-dimensional, possesses *only* the
gravitational interaction as described by Einstein's action.

Kaluza and Klein then asked how the inhabitants of this
world, too myopic to see that what they call points are actually
circles, would perceive the gravitational force. To their utter surprise, Kaluza and Klein found that the inhabitants would feel two
types of force, which they could interpret as a gravitational force,
and an electromagnetic force! In Kaluza–Klein theory, Maxwell
comes out of Einstein!

More precisely, if spacetime is really five-dimensional, then
Maxwell's electromagnetic action emerges as a piece of Einstein's
gravitational action. We can understand this stunning discovery
roughly as follows. A force in a three-dimensional space can pull
in three different directions; after all, that is what we mean by

saying that space is three-dimensional. In the four-dimensional space of Kaluza–Klein theory, gravity can pull in four different directions. To us, the myopic inhabitants, a gravitational pull in the three directions corresponding to the three directions we know and love appears as just a gravitational pull. But what about a gravitational pull in the fourth direction, the direction that we are too myopic to see? We would construe that as another force.

The reader may think of our four-dimensional spacetime as an approximate representation of five-dimensional spacetime. An action describing physics in five-dimensional spacetime splits into pieces when viewed in the four-dimensional approximation. Kaluza and Klein found that one piece describes gravity, the other, electromagnetism.

In light of the preceding discussion, the fact that two different forces would emerge from Kaluza–Klein theory is not surprising. What is surprising is that the second force has precisely the character of the electromagnetic force.

Einstein was utterly astonished. He wrote to Kaluza that the idea that space is actually four-dimensional never dawned on him. Einstein liked the theory enormously.

Figure 15.4. Dimensions of the world. (*Courtesy Topps Chewing Gum, Inc.*)

TOO SMALL FOR US TO GET IN

In Kaluza–Klein theory, the enormous disparity in strength between the gravitational and electromagnetic interactions can be accounted for if the radius of the circle is extremely small, something like 10^{18} times smaller than that of the proton. The theory replaces a fabulously small number—the strength of gravity compared to the other interactions—with another small number. At the moment, physicists have no deep understanding why, of the four dimensions of space, one is so tiny while the other three stretch clear across the universe. But the theory is consistent with observations, in the sense that the radius of the circle does not come out to be, say, one centimeter.

Kaluza–Klein theory, wild though it is, is positively tame compared to what has been imagined by science-fiction writers. No, there is no way of touring the fifth dimension. The circles are so tiny that even subnuclear particles are far too big to squeeze inside.

Over the years, Kaluza–Klein theory has inspired an assortment of crackpots to come forth with similar ideas based on unspeakable abuses of the term "dimension." The point is that it is not enough simply to assert that spacetime has however many dimensions one fancies. Kaluza and Klein had to analyze the action in detail to see what they would get in four-dimensional spacetime. Whether or not electromagnetism emerged was not up to them to decide.

THE RULE OF LOCAL SYMMETRY

The most astonishing feature of Kaluza–Klein theory, that gravity begets electromagnetism, is now understood as a consequence of local symmetry. Recall the discussion on local symmetries in Chapter 12. Einstein had built his theory on local coordinate transformation, thus inspiring Weyl also to base electromagnetism on a local symmetry, a symmetry now known as gauge symmetry. In Kaluza–Klein theory, the action written in five-dimensional spacetime possesses a local symmetry, namely invariance under five-dimensional local coordinate transformations. When spacetime is reduced to four dimensions, the local symmetry that the action possesses cannot be lost. Thus, the pieces into which the action splits cannot help but be those actions that are known to possess local symmetry, namely the Einstein action and the Maxwell–Weyl action.

The discovery of the weak and strong interactions, two interactions that apparently had nothing to do with local symmetries, consigned Kaluza–Klein theory to the aforementioned dustbins. Phenomenology ruled, and Kaluza–Klein theory, based on geometry, appeared as a hopelessly antiquated curiosity. When I studied physics, Kaluza–Klein theory was never even mentioned. The major textbooks on gravity in the 1970s did not discuss it. Then the followers of exact symmetries came roaring back. The other three interactions all turned out to be based upon the exact local symmetry of Yang and Mills. Physicists searching for a link between gravity and the grand unified interaction naturally turned to Kaluza–Klein theory. But first, they had to generalize Kaluza–Klein theory to produce the Yang–Mills action.

Kaluza and Klein supposed that each point in our three-dimensional space is actually a tiny circle. It is natural to work out what would happen if each point is actually a tiny sphere. (Note that spacetime is now six-dimensional, the sphere being a two-dimensional surface.) The Yang–Mills action, amazingly enough, pops right out! More precisely, the Einstein action for the six-dimensional spacetime splits into two pieces when viewed in four-dimensional spacetime, one piece corresponding to the four-dimensional Einstein action, the other to the Yang–Mills action.

Mathematicians call spaces that are curled up, such as the circle and the sphere, "compact spaces." In general, we could suppose each point in our three-dimensional space to be actually a tiny d-dimensional compact space (so that space is actually $[3 + d]$-dimensional and spacetime, $[4 + d]$-dimensional). Given a compact space, physicists can write down the corresponding Kaluza–Klein theory.

Mathematicians have invented all sorts of compact spaces, some shaped so oddly that we could hardly picture them. In general, each compact space is invariant under some geometrical transformation. The sphere, for instance, is invariant under rotations. Indeed, the symmetry of geometrical objects provided the original motivation for our notion of symmetry. Remarkably, the geometrical symmetry of the compact space used in Kaluza–Klein theory emerges as the local symmetry of the Yang–Mills action.

GEOMETRY INTO PHYSICS

This metamorphosis of geometrical symmetry into physical symmetry is extremely beautiful to watch, but, unfortunately, it can be appreciated fully only when cloaked in the splendor of mathematics.

Throughout this book, I have tried to convey my sense of awe at the stunning beauty of the Einstein theory of gravity and the Yang–Mills theory of the other three interactions, but the realization that one emerges out of the other can only be described as absolutely "mind-blowing," to use that overworked term.

IS GRAVITY FUNDAMENTAL?

I must caution the reader that Kaluza–Klein theory is far from established, and various other ideas remain in contention. For instance, a minority opinion holds that gravity is not a fundamental interaction at all, but merely a manifestation of the grand unified gauge interaction. In this view, the gauge interaction begets gravity, rather than the other way around. The philosophy of this approach is summarized in the aphorism: *"La lumière fut, donc la pomme a chu"* (roughly, Let there be light, so the apple may fall).

Some physicists have been critical of Kaluza–Klein theory on the grounds that it exacerbates the difficulty of renormalizing Einstein's theory of gravity. The reader can easily understand that the larger the dimension of spacetime, the more processes one has to sum up, simply because any given process has more directions in which to proceed—and the more histories one has to sum up, the less likely that sum will be sensible. As we shall see, some physicists are now thinking that this difficulty may be near resolution.

MATTER AND LIGHT

Physics books used to describe the world in terms of matter and light. Our description has become more sophisticated but the dichotomy has persisted. On one side stand the quarks and leptons, known collectively as fermions; on the other, the gauge bosons and the graviton, known collectively as bosons. For instance, in Figure 14.3 (page 237), the circles denote fermions, the arrows, bosons.

Matter is composed of fermions, while the fundamental unit of light, the photon, is the typical boson. A fermion can emit or absorb a boson, and, in the process, it can either remain unchanged or transform itself into another fermion. In this sense physicists speak of bosons acting on fermions. The shuttling of bosons back and forth between fermions produces the forces we observe.

The theories discussed thus far in this book treat bosons quite differently from fermions. In a gauge theory, the symmetry group fixes the number of gauge bosons. A theorist is free, on the other hand, to assign fermions to any representation of the symmetry group.

For example, once Georgi and Glashow decided on $SU(5)$, they were forced by group theory to have a certain number of gauge bosons. The X and Y bosons are present whether Georgi and Glashow like it or not.

Group theory alone does not determine the number of quark and lepton fields: The only requirement is that they have to fit into the representations of the group. Georgi and Glashow had to appeal to experimental observations to know that each generation contains fifteen quark and lepton fields. As I explained in the preceding chapter, the seamless fit of these fifteen fields into five-dimensional and ten-dimensional representations of $SU(5)$ provides us with a major reason for wanting to believe in grand unification.

At the moment, physicists do not understand how the Ultimate Designer chose the number of fermions. The group $SU(5)$ has a twenty-four–dimensional representation, for example. A theorist can easily imagine constructing a grand unified theory based on $SU(5)$, with twenty-four fermion fields assigned to the twenty-four–dimensional representation. The resulting universe, while quite different from the one we know, is a perfectly possible universe. Why did He choose fifteen instead of twenty-four?

Thus, the dichotomy between fermions and bosons may be expressed more sharply as questions: How did the Ultimate Designer decide on the number of fermions and which representations to put them into? Indeed, why did the Ultimate Designer include fermions at all, given that the gauge symmetry does not require them?

SYMMETRY BECOMES SUPER

To answer these questions, some theorists have argued that there must be a symmetry linking fermions to bosons, under which fermions are transformed into bosons, and vice versa. They argue for matter and light having a common origin.

In tune with our age of hyperbole, the symmetry in question has been named "supersymmetry" by its inventors. I personally regret the deflation of the word "symmetry" by implication. Inevitably, proponents of supersymmetry are sometimes referred to as superphysicists, and their field as superphysics.

Disappointingly, the original motivation of linking the known fermions to the known bosons did not work out. Supersymmetry turns out to link the known fermions to bosons yet unknown, and the known bosons to fermions yet unknown. If supersymmetry is correct, then every known particle is associated with a superpartner. Double the particles, double the fun, the enthusiasts gushed.

The sudden (hypothetical) birth of so many particles overwhelmed the registrar of particle names. The registrar, in desperation, named the superpartners of the quarks and leptons, monstrously, "squarks" and "sleptons." More endearingly, he called the superpartners of the bosons by the corresponding Italian diminutives. The photon, thus, is associated with the photino, the graviton with the gravitino, and so forth. But, then the superpartner of the W boson ends up with the unhappy name Wino!

Experimenters have failed to find any of these particles required by supersymmetry. It could be that these superpartners are so massive that they cannot be produced with the energies currently available at particle accelerators. At the moment, supersymmetry, like Yang–Mills theory during the 1950s and 1960s, is a mathematical theory in search of a world to describe.

Theorists have systematically made various existing theories supersymmetrical. For instance, the supersymmetrical version of Einstein's theory of gravity, known as supergravity, extends the theory to include the gravitino.

Supersymmetry, being broader in scope than the symmetries we have considered thus far, is consequently also more restrictive. Indeed, it is so restrictive that many supersymmetrical theories cannot be constructed in four-dimensional spacetime. One is forced by the mathematics involved to consider the theory in higher dimensional spacetime. Curiously, supersymmetry leads physicists back to Kaluza–Klein theory.

SUPERSTRINGS

Of all the recent drives toward the ultimate design, the most ambitious and revolutionary is the idea of superstrings, developed by John Schwarz, Michael Green, and others. The language of fundamental physics is quantum field theory, very sophisticated and built up over the last two hundred years or so. But while quantum field theory has reached a great level of refinement in recent years, it ultimately is based on the simple intuitive notion that particles are like tiny balls that can be represented mathematically as points. In the late 1960s, the notion developed slowly that perhaps we should construct theories whose fundamental entities are represented mathematically as line segments.

The result became known as the theory of strings. A fundamental particle is represented as a bit of vibrating string. If the bit of string is much shorter than the resolution of our detection instruments, it will look like a point particle. The remarkable feature of string theory is that by vibrating in different ways, the string can appear to us as different particles. By vibrating in a certain way, it appears as a graviton, in another, as a gauge boson. Thus, string theory holds out the promise of a truly grand unification, in which gravity is tied intrinsically to the grand unified interaction.

In the last few years, Schwarz and others have imposed supersymmetry on string theory, obtaining what is known as superstring theory. It turns out that superstring theory can only be formulated consistently in ten-dimensional spacetime. Once again, to relate superstring theory to observations, Kaluza–Klein theory must be invoked.

If we do not probe the string with too fine an instrument, superstring theory effectively reduces to a field theory containing, naturally, Einstein's theory of gravity and Yang–Mills gauge theory. Recently, in the summer of 1984 to be exact, Green and Schwarz discovered that superstring theory possesses some amazingly attractive properties. In particular, thanks to its intricate symmetry structure, the quantum version of superstring theory is renormalizable. Given that superstring theory contains Einstein's theory of gravity, Green and Schwarz may have solved the long-standing problem of renormalizing gravity.

Einstein's child is finally willing to be married to the quantum, but only as part of a larger theory. Many theorists are now working on superstring theory in feverish excitement. Others remain profoundly skeptical.

THE BAROQUE AND THE ROCOCO

There, I have taken the reader to the cutting edge of physical knowledge. At the moment, it feels as if we are living in an era when a hundred flowers bloom and a hundred schools contend. Excitement is in the air. But it remains to be seen whether any of the current ideas about the ultimate design will prove to be correct. The conservative and timid may point out that even grand unification has not been definitively established by experiment.

One disturbing sign is that while the Georgi–Glashow $SU(5)$ grand unified theory fits the observed particles and their behaviors seamlessly, these further developments all invoke hitherto unobserved particles. In art history, the baroque and the rococo followed the Renaissance. In fundamental physics, after an era of unification and simplification, we seem to have entered an era of embellishments and complexities. Recent developments tend to be increasingly complicated; in particular, superstring theory involves an enormous jump in mathematical complication.

In spite of the escalation in complication, many fundamental physicists are bullish on the future. In our limitless hubris, we are beginning to feel that we are on the threshold of really knowing His thoughts.

The details of the ongoing research in fundamental physics need not concern the lay reader. The important point is that symmetry plays a dominant role in the theories being considered currently, from grand unification to superstring. The intricacies of these theories are such that no one could possibly have constructed them by following the schema of nineteenth-century physics. Physicists have to rely on the Burning Tiger.

AS THE ANCIENTS DREAMED

In a lecture given in 1933, just before the phenomenological approach was to take hold of physics, Einstein said: "I am convinced that we can discover by means of purely mathematical constructions the concepts and the laws . . . which furnish the key to the understanding of natural phenomena. Experience may suggest the appropriate mathematical concepts, but they most certainly cannot be deduced from it. . . . In a certain sense, therefore, I hold it true that pure thought can grasp reality, as the ancients dreamed."

Recent developments appear to vindicate Einstein. Our current leap in understanding results from an insistence on aesthetic imperatives.

16

The Mind of the Creator

THE FLOW OF TIME

I have saved until last the most mysterious symmetry of them all, the symmetry of physical laws under the reversal of time. Physicists say that Nature is invariant under time reversal if the laws of Nature do not determine the arrow of time. As in our discussion of parity, I must give a precise and operational definition of time reversal invariance in order to avoid potential confusion. Make a movie of any physical process. Now play it backward. Does the process we see in the backward-playing movie contradict any physical laws? If not, physicists say that the laws governing that physical process are time reversal invariant. (This operational definition obviously puts to rest the potential misconception that time reversal invariance somehow means that we can travel backward in time.)

Notice that time reversal invariance only says that the process in the backward-playing movie, the so-called time reversed process, is possible. We watch a movie of a baseball player sliding home. Run the movie backward and we laugh: We see an imploding cloud of sand and dust converging on the player lying on the ground and lifting him up. But as far as physical laws are concerned, this process is entirely possible, though extremely improbable. As the player slides home, the molecules in the player's body transfer their momenta and energies to the molecules in the ground. If we could arrange for every molecule involved in the process to reverse its direction of motion, then the time-reversed process would in fact occur.

The direction of time is laughably evident in this example. It is amusing to think of examples for which this is not the case. Consider a silent film of a person talking: Unless we can read lips, we would have a hard time determining whether the film is being run forward or backward. (But if the speaker is Italian, say, his gestures might give the game away.)

Physicists generally believe that the arrow of time in macroscopic physical phenomena is generated by the collective behavior of the enormous numbers of particles involved. Consider the well-known example of pouring hot water gently into a glass of cold water. We all know what happens: As time goes by, the water becomes lukewarm. At the microscopic level, the molecules in the hot water are moving fast, while those in the cold move slowly. When the hot comes into contact with the cold, fast and slow molecules collide. Soon, all the molecules end up moving at some stately speed, neither fast nor slow, and the water registers lukewarm.

But the physics governing molecular collisions is in fact time reversal invariant, and any given collision can be run backward. Two molecules, both moving at a stately pace, could collide in such a way that one of them goes zipping off while the other moves off slowly.

No one, of course, has ever seen lukewarm water spontaneously separating into a cold layer and a hot layer. However, the point is, physical laws do not forbid this separation from occurring. But that the water actually would separate is extremely improbable. All the fast molecules would have to find themselves, by chance, in the top layer, say, and all the slow ones in the bottom layer. Given the huge number of molecules involved, the odds against a spontaneous separation are just staggering.

Staggeringly small, yes, but not zero. If we could watch a glass of lukewarm water long enough, far longer than the present age of the universe, we would see the water separating for just a moment into ice and hissing steam.

Because a complex macroscopic phenomenon can be reduced to various microscopic phenomena, such as the scattering of two molecules, physicists can focus their attention on microscopic processes. Since the time of Newton, physicists have relentlessly tried to run every microscopic process backward to check if Nature, at the fundamental level, knows about the direction of time. Numerous experiments have been carried out, and no one has ever directly observed a physical process that cannot be run backward. Time reversal invariance, as a result, has come to enjoy the status of a sacred principle, just as parity invariance did at one time.

But the physicists' time reversal invariant world cannot possibly be correct. We feel an arrow of time. Also, experimenters have discovered indirect evidence that under some circumstances the weak interaction mysteriously violates time reversal invariance.

THE FALL OF TIME REVERSAL INVARIANCE

I explained in Chapter 3 that after the fall of parity, shell-shocked physicists went around checking all "sacred" principles. They immediately discovered that Nature also violates charge conjugation invariance, the proposition that matter and antimatter behave in exactly the same way.

It will prove convenient in this discussion to denote parity by P, charge conjugation, C, and time reversal, T. You also may recall that after physicists discovered the breakdown of parity and charge conjugation, they found that Nature still respects the combined operation CP, in which one reflects left and right and turns particles into antiparticles at the same time. But a few years later, in 1964, experimenters discovered that Nature also violates CP once in a while in the weak interaction decay of the K meson.

Now, what does all this have to do with time reversal invariance? It turns out that a rather abstruse theorem was proven in the 1950s. The theorem states that in a world described by a relativistic quantum field theory, one may violate parity, charge conjugation, and time reversal invariances to the heart's content, if one so wishes, but one can never violate invariance under the combined operation CPT. More precisely, as a theorist, I can easily write down physical laws that violate C, P, and T separately, never minding whether these laws describe the real world . . . yet, somehow, if I take any physical process and turn it into another process by reflecting left and right, replacing particles with antiparticles, and reversing the flow of time, that transformed process also is allowed by my physical laws.

This theorem, known as the CPT theorem, surely ranks as one of the strangest and deepest theorems conceived and proven by the human mind. Since relativistic quantum field theory results from the marriage of the principle of relativistic invariance to the principle of the quantum, its pedigree is impeccable. And physicists, barring an utterly unexpected development, are extremely loath to give up the *CPT* theorem.

Given the *CPT* theorem and the observation of *CP* violation, one concludes by basic logic that *T,* namely time reversal invariance, must be violated.

In summary, physicists have strong indirect evidence that Nature violates time reversal invariance, but mysteriously enough, they have never caught Her in the act. It would be more satisfying to see the fall of time reversal invariance without having to invoke any theorem. Experimenters would like to detect an actual difference between a microscopic process and its time-reversed counterpart.

TIME AND CONSCIOUSNESS

You must remember this; / A kiss is still a kiss, / A sigh is just a sigh—/ The fundamental things apply / As time goes by.
—"As Time Goes By," Herman Hupfield

I have saved the discussion of time reversal invariance for the closing pages of *Fearful Symmetry* because I do not understand it. Neither does anybody else. As a physicist, I know what I have told you about time reversal invariance: The fundamental laws of Nature do not pick out a direction of time except in the decay of a certain subnuclear particle, blah, blah, blah. But as a conscious being, I know darn well that there *is* a direction of time. I don't care what physicists say, I know that the flow of time is irrevocable. For lovers and nonlovers alike, time goes by.

In physics, time is simply treated as a mathematical parameter; as time changes, various physical quantities change in accordance with various physical laws. Einstein's work deepened the mystery by treating time and space on equal footing. Yet, again as a conscious being, I *know* that time is different from space: I can go east or west, as I please, but I can only go in one direction in time.

We are confronted here with an impasse enforced by a fundamental guiding tenet of science: the exclusion of consciousness. Physicists are careful to say that their knowledge is limited to the physical world. The realization that the world may be divided into the physical and, for lack of a better term, the nonphysical surely ranks as a major turning point in intellectual history, and one that has made possible the advent of Western science. But eventually we will have to cross the dividing line. I believe that a deep understanding of time reversal invariance will take us across that line.

Is the arrow of time perceived by our human consciousness generated in the same way that an apparent arrow of time is generated when we mix hot and cold water? Will someone someday suddenly perceive, for an instant, time flowing backward? Somehow, I do not think so. I have no good reason for believing this, but I refuse to believe that our perception of time is merely a probabilistic illusion.

The possibility that our perception of time is linked to the violation of time reversal invariance in the decay of some subnuclear particle appears untenable, if not absolutely inconceivable. K mesons, surely, are not present in our brains. Besides, how can a tiny effect in the weak interaction govern the overall working of the mind, which, as some would have it, is entirely driven by the electromagnetic interaction? Physics has not been able to provide any answers.

That there is consciousness in the universe is undeniable. That science in general, and physics in particular, do not address this most striking of all observable phenomena is glaring. Consciousness, so central to our existence, remains a mystery.

A tantalizing clue comes from quantum physics. Ever since the early days of the quantum, when it was realized that the act of observation unavoidably disturbs the observed (as quantified by the uncertainty principle), physicists and philosophers have speculated about the possible link between consciousness and the probabilistic mystery of the quantum. There is no lack of speculation and musing on the subject, but it is fair to say that the overwhelming majority of working physicists find what has been written exceedingly difficult, if not impossible, to understand. The distinguished physicist Murph Goldberger was once asked by a television interviewer why he had never worked in this area. He answered that every time he decided to think about these questions, he would sit down, get out a clean piece of paper, sharpen his pencil—and then he just couldn't think of anything to put down. That is as good a summary as any of our present understanding of the role of consciousness in physics.

Ultimately, the discussion comes down to the question of whether science can explain life; that is, whether there is a "life force," for lack of a better term, outside the purview of rational thought. Is the human consciousness merely the result of a bunch of neurons exchanging electromagnetic pulses? Is the thinking brain ultimately just a collection of quarks, gluons, and leptons? I don't think so. Do I have a cogent reason? No, it is just that, as a physicist, I do not have enough hubris to believe that physics can be all encompassing. When He set down the symmetrical action of the world, did He see human consciousness in it? Is consciousness a piece of the action, or is it beyond the purview of a symmetrical action? Sir Arthur Eddington (1882–1944), a distinguished English astrophysicist who was beset by bizarre ideas toward the end of his life, once gave the following parable: In a seaside village, a fisherman with a rather scientific bent proposed as a law of the sea that all fish are longer than one inch. But he failed to realize that the nets used in the village were all of a one-inch mesh. Are we filtering physical reality? Can we catch consciousness with the nets we are using?

Such are the night thoughts of a fully conscious, contemporary physicist who's a bit afraid of the dark. But I better stop and go back to something I believe in: symmetry, for instance.

THE NATURE OF THE BEAST

We have traveled a long road with the Burning Tiger. Starting with the discovery that heaven is not above us, symmetries have played an increasingly central role in our understanding of the physical world. From rotational symmetry, physicists went on to formulate ever more abstruse symmetries. But the basic notion and motivation remain the same. Fundamental physicists are sustained by the faith that the ultimate design is suffused with symmetries.

Contemporary physics would not have been possible without symmetries to guide us. Einstein showed us how the secrets of gravity could be mastered in one fell swoop. Learning from Einstein, physicists impose symmetries and see that a unified conception of the physical world may be possible. They hear symmetries whispered in their ears. As physics moves further away from everyday experience and closer to the mind of the Ultimate Designer, our minds are trained away from their familiar moorings. We need the Burning Tiger.

Writing about Blake's poem on the Burning Tiger, noted critic Lionel Trilling pointed out that, up until the fifth stanza, the poet sought to define the nature of Tiger by the nature of God, but that in the sixth and last stanza, the tone of the poem shifts, and it is God who is defined by the nature of the Tiger. In the same vein, I like to think of an Ultimate Designer defined by Symmetry, a *Deus Congruentiae*.

The point to appreciate is that contemporary theories, such as grand unification or superstring, have such rich and intricate mathematical structures that physicists must marshal the full force of symmetry to construct them. They cannot be dreamed up out of the blue. Nor can they be constructed by laboriously fitting one experimental fact after another. These theories are dictated by Symmetry.

Do recent developments in fundamental physics represent the beginning of the end in our search for understanding, or do they merely signal the end of the beginning? The optimists proclaim that we will know the ultimate design any day now. The pessimists mumble that we are like jigsaw-puzzle solvers who succeed in fitting four pieces together, not realizing that hundreds more remain in the box. The traditionalists decry basing theories on aesthetics rather than on cold hard facts. Certain theories, such as superstring, are so far removed from perceived realities that they could well come crashing down. Or perhaps they will become sterile, as did atomic theory, for example, in the time of Democritus.

DID HE HAVE ANY CHOICE?

When judging a scientific theory, his own or another's, he asked himself whether he would have made the universe in this way had he been God. This criterion ... reveals Einstein's faith in an ultimate simplicity and beauty in the universe. Only a man with a profound religious and artistic conviction that beauty was there, waiting to be discovered, could have constructed theories whose most striking attribute, quite overtopping their spectacular successes, was their beauty.
—Banesh Hoffman

It has been said that the highest praise of God consists in the denial of Him by the atheist, who finds creation so perfect that he can dispense with a creator.
—Marcel Proust

As fundamental physicists drive toward the ultimate design, they begin to confront the issue of uniqueness. If we believe the ultimate design to be the most beautiful possible, does it follow that it is the only one possible? Einstein once said, "What I'm really interested in is whether God could have made the world in a different way; that is, whether the necessity of logical simplicity leaves any freedom at all." Most of us who work on fundamental physics share this sentiment. We want to know if He had any choice.

Grand unified theories fail the test of uniqueness. In constructing one, we can choose any group we please, and once the group is decided upon, we can choose the representations to which the fermions belong. Why would He choose $SU(5)$, if indeed He did? Why didn't He choose $SU(4)$ or $SU(6)$, or $SU(497)$, for that matter, if indeed He didn't? We do not know why. Of course, most choices would not lead to the world as we know it, but that's beside the point.

Some theorists amuse themselves by posing the following sort of questions, "If I were given forty-five fermion fields, how would I design the universe? Would I choose $SU(5)$, and put the fermions into three generations, or is there a better design?" Remarkably, by imposing only a few rules based on general principles, the choices can be narrowed down considerably. But who decides that there are to be forty-five fermion fields?

Many fundamental physicists believe that by imposing ever more stringent symmetries, we may find that we have only one choice for the action of the world. While philosophers such as Pangloss once attempted to demonstrate that ours is "the best of all possible worlds," fundamental physicists are now smitten with the ultimate hubris of wanting to prove that ours is the *only* possible world. (Actually, this view is not entirely novel. Already, in the early eighteenth century, Leibniz, puzzling over why the world is the way it is, felt that God must have had a good reason for creating our particular world rather than some other out of the infinitude of possible worlds.)

The reader should not confuse this desire to prove that God had no choice with a class of arguments known as the "anthropic." They purport to show that the world must be the way it is because, otherwise, intelligent beings, such as we humans, would not have been possible.

As an example, consider the burning of stars. The strong

force between two protons inside a star tends to push the two protons together, while the electric force between them tends to pull them apart. As I discussed in Chapter 2, were the strong force just a wee bit stronger, the two protons would come together rapidly, releasing energy. Stars would soon burn out, making steady stellar evolution and biological evolution impossible.

Proponents of the anthropic argument point to this intricate balance and exclaim that the existence of intelligent beings requires that the strong interaction cannot be stronger than a certain amount. They then try to find another situation in which the comfort of intelligent beings would be seriously compromised, were the strong interaction weaker than it actually is. They hope to show, in this way, that the fundamental laws of physics must be as they are.

The trouble with anthropic arguments is that they only show that to support life as we know it, the world has to be intricately balanced. The world is the way it is because it is the way it is. While anthropic arguments are often very interesting, many physicists, including myself, do not find them intellectually satisfying.

In the anthropic view, the Ultimate Designer is a tinkerer. He tried out one design after another until He found one that accommodates intelligent beings. Living in this age of computer-aided design, when an engineer can try out many different designs by simply pushing a few buttons, I can even imagine that He actually created an infinite number of universes, one for each of the infinitely many possible choices of groups and representations. Hmm, this universe, based on $SU(4)$, doesn't work too well. The one based on $SU(6)$, which I tried yesterday, was even worse. But hey, look at this one, based on $SU(5)$; looks like it will make a fun universe.

Like many of my colleagues in fundamental physics, I prefer Einstein's view. I like to think that a truly great architect, when shown a site and a program, would proclaim instantly that there is only one design possible. Surely, He too was driven by the implacable force of aesthetics to a unique design.

At the moment, this somewhat mystical view that God had no choice remains little more than a cherished dream. We are still groping to discover what the relevant aesthetic criteria are, but we do not doubt that symmetry will light our way in our quest to know His mind.

Afterword

In sitting down to write this Afterword to the Princeton University Press edition, I realized that it should consist of two parts: the first a more or less straightforward journalistic reporting on what has happened in fundamental physics in the thirteen years since *Fearful Symmetry* was first published, the second an elaboration of the theme of symmetry in the ultimate design that runs through this book.

SUPERSTRING

Without question, superstring theory represents by far the most significant development in fundamental physics since *Fearful Symmetry* was published. But to explain superstring theory clearly in a limited amount of space is literally impossible. An entire book is needed, and entire books have been written.

I did, however, touch upon superstring theory in Chapter 15. I finished writing *Fearful Symmetry* just as the superstring revolution swept through the theoretical physics community. Within the limited space available here I will do the best I can to give the reader a little bit more. My solution is to include in an appendix a chapter titled "The Music of Strings" taken from my book *An Old Man's Toy,* published in 1989. In it I told about my personal encounter with the dramatic birth of modern superstring theory in 1984. I described the spectacular triumphs (referred to as "miracles" by the true believers) of the theory, as well as its extravagant excesses. The interested reader is urged to turn to the appendix now.

Upon re-reading "The Music of Strings" the other day I marveled that as an introduction to superstring theory for the general public, it is still more or less adequate after ten years. The deep rift in the fundamental physics community over superstrings that I described is still with us, a rift perhaps deeper than ever, reflecting profound differences in philosophy over how fundamental physics should proceed. In "The Music of Strings" I mentioned that, to develop superstring theory, physicists had to learn an enormous amount of fascinating and difficult mathematics. This is perhaps not surprising considering how fundamentally new the conceptual foundation of the theory is: superstring theory is the most conceptually novel advance in theoretical physics since the invention of quantum field theory. Indeed, superstring theory has impacted pure mathematics, triggering perhaps the most dramatic infusion of new ideas into pure mathematics in decades. The leading superstring theorist and most vocal prophet of the theory, Edward Witten, was awarded the Fields medal, the equivalent of the Nobel Prize in the mathematical community.

The road superstring theory has traveled during this last decade and a half has been bumpy and arduous. The theory went through enormous upheavals. Various developments inspired feverish excitement in the superstring community. A mere few years later these developments were regarded as largely irrelevant. As the eighties turned into the nineties, discouraged people left the field in droves, but then in the mid-nineties the theory came roaring back. Branes were discovered.

What are branes, you cry, I thought we are talking about strings!

In the appendix, I explained that physicists constructed all of physics on the notion that the fundamental particles, such as the electron and the quarks, can be represented as mathematical points. The conceptual advance of string theory consists of saying that the fundamental particles are actually wriggling pieces of strings. The idea is that the pieces of strings are so tiny that unless we have extraordinarily fine instruments we would have thought that they are point particles. By wriggling in different ways, the string would appear as the various fundamental particles, previously thought of as points.

BRANES AND BRAINS

A word about terminology: In the language of mathematics, point particles are zero-dimensional objects, while strings are one-dimensional objects. Thus, with string theory fundamental physicists have gone from studying zero-dimensional objects to studying one-dimensional objects.

You the reader might be wondering: well, by that same line of reasoning, why not imagine the world as made up of two-dimensional objects?

That was a brilliant remark. Physicists have indeed imagined this possibility. Think of a long hollow tube. It is a two-dimensional object. (When counting dimensions, mathematicians look at the surface of the tube, not the space enclosed by the tube. The surface of a long hollow tube is two dimensional.) If the tube is much longer than its radius and if you are looking at it from far away, then the tube would appear to you as just a string, that is, a one-dimensional object. So, in the same way that a loop of string looked at from far away would look like a point, a tube looked at from far away would look like a string.

Physicists refer to two-dimensional objects, that is, surfaces, as membranes. This is of course just part of the jargon; you should not be thinking about the biological membranes in your body.

So why not construct a theory beyond the theory of strings? Why not construct a theory of membranes? In fact, why not keep on going, and consider a theory based on three-dimensional objects, which the more playful among theoretical physicists refer to as blobs, and the more academic as three-dimensional membranes?

Hey, we all know how to count: after three comes four, and after four comes five. Why not a theory based on four-dimensional membranes? And so on and so on. It doesn't take a genius to see that, in principle, we could be talking about a theory based on p-dimensional membranes, with the letter p standing for any integer, seven, or twelve, or whatever you like.

Naturally, theoretical physicists soon abbreviated the unwieldy term p-dimensional membranes to p-branes, prompting some jokester to quip that to work on the theory of p-branes you've got to have a pea brain.

Thus, as soon as theoretical physicists try to flutter away from the point particles that their subject was founded on, they are logically compelled to consider ever higher dimensional objects. A Pandora's box has been opened, and the foes of string theory are outraged.

It turns out that the construction of a quantum theory of membranes from scratch is beyond the present mathematical capability of theoretical physicists. That is, of course, not an argument that the fundamental theory of the world may in fact not be based on membranes. It is just that string theorists did not know how to deal with all these p-branes.

Thus, for many years, the leaders in superstring theory adopted a pragmatic stance. They simply decreed that, for the moment, they would consider only strings.

Later, in a rather ironic turnaround, string theorists showed that the mathematical consistency of superstring theory requires the presence of the p-branes. You can't keep the p-branes out. Suppose you start out to construct a theory based only on strings, you will eventually discover that for the pieces of the theory to fit together mathematically, you are forced to include the p-branes.

The subject is sometimes called brane physics, and inevitably some wit has started referring to the kind of physics done by all other physicists as brane-less physics. The confusion of brane and brain in conversation is cause for hilarity. In seminars, people speak of "your brane" and "my brane." It does not take much to get theoretical physicists to laugh. Lately, the joke has gotten a bit old and the hilarity has died down somewhat.

THE BIG BRANE

In an extremely amusing twist, some people now speculate that certain branes can be big, very big.

You learned in school that particles such as the electron are tiny, a zillionth of a centimeter in size. As I explained in "The Music of Strings," the loop of string whose shaking and rolling produces the electron has to be correspondingly tiny. Originally, people naturally thought that membranes and p-branes also have to be small.

But, remember that a three-brane is just a three-dimensional mathematical structure. Thus, our entire three-dimensional universe could very well be a three-brane sitting in a nine-dimensional space. (In "The Music of Strings" I mentioned that string theory is formulated in nine-dimensional space or ten-dimensional spacetime.) In other words, we could be living in a three-brane. A very big three-brane! Our universe is a three-brane. (Note, by the way, that here we are just talking the spatial structure of the universe; time is an extra dimension we are implicitly keeping in the back of our minds. There is no mystery to this; physicists speak of our universe as either three-dimensional or four-dimensional, and it is understood that in one case time is counted and in the other it is not.)

In fact, the wild notion that we are living in a three-brane sitting in a nine-dimensional space can be developed considerably further. Recall that in Chapter 14 "Unity of Forces" I described a grand unified theory based on the group SU(5). Remember what the math talk meant? There are five fields that transform into each other under the group SU(5). Let us see how this could emerge in the brane theory.

We human beings were equipped by evolution to be able to picture three-dimensional space. Most of us have a hard time visualizing three-branes sitting in a nine-dimensional space. To convey to you the general idea, let me talk about two-branes sitting in a three-dimensional space. The three-dimensional space is easy to visualize: it is just the ordinary everyday space we live in. The two-branes, being two-dimensional, can be thought of as sheets of paper. Each of the two-branes represents a two-dimensional universe; this is the analog of the wild notion mentioned above that our universe may be a very big three-brane.

Now, let us consider a situation in which five two-branes lie on top of each other, in exactly the same way that five sheets of paper lie on top of each other in a stack. (A sheet of real paper of course has a finite amount of thickness, no matter that it is the thinnest we can find in a stationery store. Mathematically, we think of the two-branes as having no thickness at all, so that the five two-branes form one single two-dimensional universe.) On each of the two-branes there are physical excitations represented by quantum fields. Let us focus on a particular field on a two-brane. In this two-dimensional universe, there are thus five fields, distinguished by which two-brane each of them is associated with. Voilà! This could be the origin of the SU(5) grand unification group we talked about in Chapter 14, on which all known physics may be based. (The discussion works the same way for the "actual" case of five three-branes sitting on top of each other in a nine-dimensional space.)

THE ELUSIVE M THEORY

Soon after the realization that branes could not be excluded, another far-reaching development rocked the superstring community. In the initial euphoria over the "three miracles" of superstring theory, the faithful thought they had found the one true theory of the world. Now, if you really feel that you have found the Holy Grail, you don't want to discover that there are actually five different holy grails. But that was exactly what happened with superstring theory. People discovered that there was not one superstring theory, but actually five different theories. Depression set in in certain quarters of the super-string community, while their foes snickered. How can the theory be the real thing if there are five of them?

Eventually, superstring theorists realized that the five different theories are actually different facets of one single mysterious theory that they cannot even begin to write down. Ed Witten dubbed this all-mysterious theory M theory. When asked what the letter M stands for, Ed said slyly that it could stand for membrane, mystery, magic, or mother. Some wit (Is this always the same guy?) promptly referred to the new theory as "the theory formerly known as string theory," alluding to a well-known entertainer in popular American music who changed his name to revitalize his career.

The situation is literally like the old fable about a group of blind men who tried to find out what an elephant looks like. One blind man touched the legs and asserted that an elephant is like a tree, while another touched the trunk and insisted that an elephant is like a rope, and so on. M theory is the elephant. The five different superstring theories are akin to the five different descriptions of the elephant given by five blind men. Nobody knows what M theory looks like.

Perhaps each of the five different superstring theories describes one aspect of M theory, that mysterious mother of all theories. In an even more amazing twist, a quantum field theory, known to the con-noisseurs as eleven-dimensional supergravity but, in spite of the fancy name just a garden variety quantum field theory of the type described in much of this book, also turns out to describe one aspect of M the-ory. Wow, a plain old field theory has also joined the dance, just as dazzling as the five superstring theories, and having just as much right to be in the dance as they. After superstring theorists had already dis-missed field theories as some inferior poor cousins to superstring theo-ries, a strange eleven-dimensional field theory has come back. What a dramatic turn of events!

What is going on? Nobody, but nobody, knows. Meanwhile, the deep division between theoretical physicists described in "The Music of Strings" persists.

The skeptics continue to express grave doubts, for some of the reasons I mentioned. They believe that superstring theory will be swept into the dustbins of physics history, that in the end its impact will prove to be mainly on mathematics. It could well be that, decades from now, physics historians will look back at a strange episode during which some physicists believed in a momentary aberration known as superstring theory. There are silent skeptics and vocal skeptics. The silent skeptics simply decline to work on superstring theory and try to find something else in theoretical physics that will challenge their minds. The vocal skeptics, on the other hand, actively snipe at superstring theory. I personally feel that the vocal skeptics, and there are fewer and fewer of them as the years go by, should be quiet and leave the superstring theorists alone, particularly since they have nothing to offer as a replacement for superstring theory. The lay person sometimes gets the impression that in theoretical physics it is one theory against another. But in fundamental physics, it is often one theory against nothing, or nothing structured enough to be called a theory by any reasonable definition.

On the other side, the faithful sometimes sound like they are on some kind of jihad, so sure are they that they are on the right path. Their deeply held faith that superstring theory is correct, a faith with an almost religious fervor to it, is partially justified and certainly beguiling. Superstring theory and, by extension, M theory, whatever it is, is incredibly beautiful, surpassing anything theoretical physicists have ever laid eyes on before. But part of it is also simply driven by the sociology of academia, unfortunately. Imagine you have devoted years as a graduate student to mastering superstring theory and then have struggled to find a postdoctoral position, and then another postdoctoral position. After an enormous fight, you have managed to beat out all of your superstring colleagues at the postdoctoral level to obtain an assistant professorship at a leading university where superstring theory is thriving. Do you think, after all that, that you can suddenly come out one day and announce that you no longer believe in superstring theory? It takes much strength of character.

What I think is of no importance to the string theorists. But my opinion may be of some interest to the reader. I, too, am dazzled by the mathematical beauty of superstring theory and would like to believe that the superstring theorists are on the right track. Nevertheless, I have an uneasy feeling that the theory is too conservative.

That seems like a strange thing to say about a theory generally billed as truly revolutionary. Let me explain. When physicists started to explore distance scales about the size of an atom, they found that classical physics was no longer valid, and a radically different physics had to be developed. Thus was quantum physics born. As I explained in Chapter 10, the basic notions of classical physics, such as position and momentum, can no longer be defined. The entire framework in which theoretical physicists operate has to be changed.

In contrast, in string theory, once the superstring theorists replaced the notion of point particles by that of strings (and even there, the string can be loosely thought of as made of particles joined together) they use many of the same concepts described in this book. They write down an action and apply the standard principles of quantum physics. Many physicists are reluctant to believe that in going from the size of a proton to the size of string, traversing a factor of something like 10^{19} in distance, the standard formulation of quantum physics would continue to be valid.

THE NECESSITY FOR SYMMETRY?

In this book, we learned that symmetry is an indispensable notion used by theoretical physicists to organize and thus simplify an otherwise bewildering array of phenomena in the unfamiliar arena of the microscopically small. In Chapter 16, I told the fisherman's parable of Sir Arthur Eddington. In a fishing village, a fisherman somewhat inclined towards discovering scientific laws is convinced that he has found a fundamental law of the sea, that all fish are longer than one inch. He has somehow failed to notice that the fishing nets used in the village all have one inch meshes.

In sifting and organizing the tidal wave of experimental data, Eddington asked, might the minds of the theoretical physicists automatically cast back into the sea those theories that are not symmetric?

We can well ask about the unreasonable effectiveness of symmetry in our understanding of Nature. Is the Ultimate Design truly symmetric, or is it that we can only understand the symmetric part of the Ultimate Design?

An exciting new field of evolutionary psychology has emerged in recent decades through the work of Leda Cosmides, John Tooby, and others. Just as evolution (or better, natural selection) has inexorably, over a hundred millennia, molded human biology, evolution surely must also have shaped the workings of the human mind. A considerable body of experimental evidence now supports this perfectly plausible hypothesis. The human mind is not an all-purpose computer, as some earlier thinkers suggested, but consists of various interlinked "modules" dedicated to specialized computation. For example, it has been shown experimentally and quantitatively that the human mind contains a module dedicated to detecting symmetry or the lack thereof in the bodily features of potential mates, an ability that clearly confers a reproductive advantage.

The mind of the theoretical physicist is also trained, or perhaps naturally inclined, to see symmetries everywhere. A psychologist friend of mine wryly remarked to me that one of the symptoms used to diagnose obsessive compulsive disorder is an excessive interest in symmetry and order.

DEISM VERSUS THEISM

This book is partly inspired by Einstein's remark that what he wanted was "to know how God created this world, . . . to know His thoughts," as I quoted at the beginning of Chapter 2. Throughout the book I spoke freely, and as a writer I enjoyed speaking freely, of God, the Creator, the Ultimate Designer, Mother Nature, and various similar and interchangeable names for presumably the same Entity. Naturally, some readers (and some book reviewers) asked me about my thoughts on "the Ultimate Designer." My thoughts, for what they are worth, incline towards the deistic rather than the theistic. Deism holds that the universe is created by some all-encompassing "Presence," while theism asserts that the Creator, whoever He, She, or It might be, takes an active interest in the pains, joys, and aspirations of individual humans, or perhaps in the spirit of political correctness, of all sentient beings in the universe. Nothing in this book implies or necessitates the theistic view.

I was quite pleased when *The New York Times Book Review* declared that I "conveyed the religious beliefs of theoretical physics with far more accuracy than, for example, Fritjof Capra did in *The Tao of Physics*." They said the religious beliefs of theoretical physics, but perhaps they should have said the religious beliefs of the theoretical physics community. My impression is that the religious beliefs of this community (more accurately the American segment of this community which I happen to know best) range over the entire spectrum, from the militantly atheistic to the deeply devout, with the distribution dropping off sharply towards the devout end. I think that many theoretical physicists are awed by the elegant structure that underlies fundamental physics. Those who have thought about it are struck dumb with astonishment, as was Einstein, that the world is in fact comprehensible.

I think this dramatic realization—that the world is comprehensible—is one of the most profound insights in intellectual history. We only have to recall that many splendid and sophisticated civilizations, notably some in the Orient, never presumed that this was so.

Contemporary evolutionary biologists have reached the conclusion, which strikes me as convincing, that there is no design in the biological world or, to put it more accurately, that the apparent design in the biological world can be understood as arising through random mutation and natural selection. This point of view was most forcefully stated by Richard Dawkins in his book *The Blind Watchmaker.* To me, the strongest argument against creationism consists in pointing to all the bad designs in our bodies, as documented recently with care and elegance by George C. Williams in *The Pony Fish's Glow.*

Of course, if there is no design in the biological world, this does not logically imply that there is no design in the physical laws that ultimately govern the biological world. The Ultimate Designer could be considerably more subtle than the eighteenth-century theologian William Paley and his followers. (Here I recall yet another famous quote attributed to the Grand Old Man of physics: "Subtle is the Lord, but malicious He is not.") He (that's a capital H) in all likelihood did not care about whether the testis in the human male would drape over the ureter in their evolutionary descent towards the scrotum, but He might have taken care to pick the right gauge group so that the laws of physics would lead to stars and planets and the laws of biological evolution and ultimately to a species of apes, perhaps among many many intelligent beings in many galaxies, who could ponder whether they are deistic, theistic, atheistic, or whatever. That to me would indicate a subtle sense of humor.

Funny is the Lord, but benevolent He is not.

THE ULTIMATE DESIGN?

As I suggested in Chapter 16, perhaps the Ultimate Designer is not that clever, perhaps He is the Ultimate Tinkerer, who tried out one design after another until He found one that would produce intelligent beings capable of debating about His existence. He tried one group after another. Hey, SO(7) doesn't work, let's try SO(10) instead. Perhaps there are an infinite number of universes, one for each choice of the fundamental theory, some having an interesting theory, but with most collapsing almost immediately.

Indeed, this kind of thought has been carried to the extreme by the relativist Lee Smolin, who tried to combine Darwin and Einstein, as it were, by imposing the idea of natural selection on cosmology. Only those universes with the most interesting physical laws will survive to sire more universes. The idea, while amusing, strikes most physicists (not to mention biologists!) as stretching Darwin's theory quite a bit. For one thing, universes presumably do not mate to exchange "genetic material."

Along this line, an amusing science fiction scenario could be developed by imagining the universe as a school-assigned science experiment carried out by a high-school student in a meta-universe. Perhaps he or she or it has even started a whole bunch of universes, like ant farms, and stashed them away somewhere in the basement, out of his or her parent's way. Perhaps he (that's a lower case H) has lost interest and forgotten about his universes, leaving some to expand, others to collapse, in complete futility and silence.

I had better not develop this scenario any further, lest this book not be publishable by a prestigious university press.

Appendix to Chapter 9

For the mathematically inclined reader, I will indicate why the nine entities obtained by gluing two defining representations of $SO(3)$ together split up into separate clans.

Let's look at the nine: ⓑ Ⓡ, ⓡ Ⓨ, ⓨ Ⓡ, and so on. Select any one entity and interchange the colors indicated in the circle and in the square. The entity ⓡ Ⓡ does not change, but the entity ⓡ Ⓨ becomes ⓨ Ⓡ, and the entity ⓨ Ⓡ becomes ⓡ Ⓨ. Mathematicians had the key idea of dealing with the linear combinations ⓡ Ⓨ + ⓨ Ⓡ and ⓡ Ⓨ − ⓨ Ⓡ, instead of ⓡ Ⓨ and ⓨ Ⓡ.

Why in the world is this a good idea? The point is, when we interchange the colors indicated in the circle and the square, the combination ⓡ Ⓨ − ⓨ Ⓡ, which mathematicians call odd, becomes ⓨ Ⓡ − ⓡ Ⓨ, which is equal to − (ⓡ Ⓨ − ⓨ Ⓡ). The odd combination, in other words, becomes minus itself. In contrast, the combination ⓡ Ⓨ + ⓨ Ⓡ, which mathematicians call even, becomes ⓨ Ⓡ + ⓡ Ⓨ, which is just itself. It does not change, in other words. The entities ⓡ Ⓡ, ⓨ Ⓨ, and ⓑ Ⓑ are also even in this sense: When the colors indicated in the circle and square are interchanged, these entities do not change.

We have managed, thus, to separate the nine entities into two clans: one that includes three odd combinations, the other, six even ones. It is clear that there are three odd combinations; in fact, we can easily list them: ⓡ Ⓨ − ⓨ Ⓡ, ⓨ Ⓑ − ⓑ Ⓨ, and ⓑ Ⓡ − ⓡ Ⓑ. Similarly, we can list the even combinations.

Good. Now we must examine how these combinations are transformed by a rotation.

Consider a rotation that takes the arrow $\vec{x}$ into $a\vec{x} + b\vec{y} + c\vec{z}$. (As explained in the text, a, b, and c are just good old ordinary plain numbers.)

Then, the entity ® ⃞R̲ is transformed into the linear combination:

$$(a \text{ ®} + b \text{ Ⓨ} + c \text{ Ⓑ})(a \text{ R̲} + b \text{ Y̲} + c \text{ B̲})$$
$$= a^2 \text{ ®} \text{ R̲} + ab \text{ ®} \text{ Y̲}$$
$$+ ac \text{ ®} \text{ B̲} + ba \text{ Ⓨ} \text{ R̲}$$
$$+ b^2 \text{ Ⓨ} \text{ Y̲} + bc \text{ Ⓨ} \text{ B̲}$$
$$+ ca \text{ Ⓑ} \text{ R̲} + cb \text{ Ⓑ} \text{ Y̲} + c^2 \text{ Ⓑ} \text{ B̲}$$

Nothing fancy was done here. We merely "multiplied out" the expression $(a \text{ ®} + b \text{ Ⓨ} + c \text{ Ⓑ})(a \text{ R̲} + b \text{ Y̲} + c \text{ B̲})$. Thus, the first term, $a^2 \text{ ®} \text{ R̲}$, comes from multiplying $a \text{ ®}$ by $a \text{ R̲}$, the second term, $ab \text{ ®} \text{ Y̲}$, from multiplying $a \text{ ®}$ by $b \text{ Y̲}$, and so on.

All right, you say, you've just shown me that the simple-looking entity ® R̲ is transformed by a rotation into a mess. Now what?

Luckily, we don't really have to tackle that messy-looking combination. All we have to do is notice that it contains only even combinations. Thus, for example, the entity ® Y̲ appears multiplied by the number ab, and the entity Ⓨ R̲ appears multiplied by the number ba, which, of course, is the same as ab (a and b denote numbers). In other words, the even combination ® Y̲ + Ⓨ R̲ appears, but not the odd, ® Y̲ − Ⓨ R̲.

Wait, this is totally obvious! We are looking at the transformation of an even combination, ® R̲. For an odd combination to appear, a minus sign has to appear. But a minus sign cannot just pop out of thin air!

There, we've done it. We have just proved that even combinations transform only into even combinations. Similarly, odd combinations transform only into odd. The division of even and odd combinations conveys the essence of why a representation obtained by gluing two representations together will, in general, split up into smaller representations.

Instead of continuing with an exhaustive (and exhausting) mathematical analysis, I content myself with having given you a flavor of the argument involved. I will not go on, therefore, to show why the six even combinations split up further into a clan of five combinations and a clan of one combination.

To sum up, let us refer to our silly analogy. The extraterrestrial realized that he should classify objects by how many legs they have. Since the pumpkin has no legs, it cannot be transformed into the prince. Here, we are more sophisticated: We classify the combinations by whether or not they have a minus sign.

You have now learned more group theory than you may realize. For instance, in Chapter 13, I mention that the Ultimate Designer used the rotation group in ten-dimensional space, $SO(10)$. Well, now you can actually work out how to glue two copies of the defining representation of $SO(10)$ together. There are presently ten possible colors and one hundred entities in 10×10: ⑧ Ⓡ, ⑧ Ⓨ, and so forth. Well, how many odd combinations are there? Let's count. We can choose ten possible colors to put in the circle. After coloring the circle, we are left with nine possible colors for the square (since we cannot choose the same color for the circle and the square in an odd combination: ⑧ Ⓡ − ⑧ Ⓡ − 0!).

Furthermore, since ⑧ Ⓨ and Ⓨ Ⓡ appear together in one combination, we should divide by two to avoid counting twice. Thus, we have altogether $(10 \times 9) / 2 = 45$ odd combinations. Using this fact, physicists deduced that if the world is really described by a gauge theory based on $SO(10)$, then there must be forty-five gauge bosons. (Gauge theory and gauge bosons are discussed in Chapter 12.)

Of the $100 - 45 = 55$ symmetric combinations, one transforms into itself. (I do not explain this mathematical fact here.) Thus, in $SO(10)$ we have: $10 \otimes 10 = 1 \oplus 45 \oplus 54$. (See, group theory is not *that* difficult.)

Appendix to the Afterword

A CABARET SHOW

In the summer of 1984, my wife and I drove across the country with our children, stopping at various physics centers along the way, much like migratory wildebeests seeking out watering holes. At Aspen, Colorado, we stopped for a month. There, amid the splendors of the Rocky Mountains, physicists gather every summer to discuss the latest sense and nonsense. The atmosphere at Aspen always has been rather relaxed: The physics chitchat is mixed in with volleyball and picnics and hikes and music and the local scene, such as it is.

One warm summer's eve, as part of the fun and frivolity, the physicists took over the bar at the venerable Hotel Jerome to stage a variety show. In one of the skits, a physicist came out ranting and raving that at long last he understood the secrets of gravity. He now had a theory of the whole world, including gravity. Before he could explain what it was, however, two men dressed in white coats came and dragged him away.

The skit is a familiar one, and the sight of a mad physicist being dragged off to an asylum can always be counted on to provoke a wave of titters among a physics audience. Sitting in the audience, I felt that there was a certain Gary Larson flavor to the whole scene. But that summer, the physicist playing the mad physicist was perfectly serious about having a theory of the whole world.

Some days earlier, when I first arrived at the Aspen Center for Physics, I ran straight into John Schwarz, who played the mad physicist, and his collaborator Mike Green. They were both excited. Mike Green and I had been postdoctoral fellows together, and I know well how he talks

This appendix first appeared as Chapter 13, "The Music of Strings," in *An Old Man's Toy: Gravity at Work and Play in Einstein's Universe*, by A. Zee, Macmillan Publishing Company, New York, 1989.

when he gets excited. Green and Schwarz had been working on something called the string theory, later to become the superstring theory, on and off for a decade, and now finally they had a version of the theory that worked.

NOT INFECTED WITH ANOMALY

When physicists write down a theory of the world, they must check whether the theory can be combined with the quantum principle. In the 1960s, physicists discovered that even though some theories may look perfectly sensible at first, they can harbor a mathematical inconsistency called the *anomaly*. If your favorite theory suffers from the anomaly, then it cannot be combined with the quantum principle and you can kiss it good-bye. To check whether a theory is infected with the anomaly, physicists developed what amounts to a clinical test. You look at your theory and calculate a number according to some formula. That number determines the degree to which the theory is infected with the anomaly. If you get zero, the theory is not infected. If you don't get zero, it goes into the trashbasket.

Superstring theory is notorious for being afflicted with the anomaly, and that partly accounts for the general lack of interest in the superstring theory until that summer. Green and Schwarz had been computing the degree to which various versions of the superstring theory were afflicted with the anomaly. Mmm, not zero. No good. Try another one. Finally they found a version in which the number, remarkably enough, happens to be zero. Hence their enormous excitement when I ran into them. They showed me on the blackboard how the numbers precisely cancel. I was particularly interested, since throughout my career I had worked on the anomaly from time to time. I was astonished by the almost magical way in which the anomaly in the Green and Schwarz version of the superstring theory cancels to zero. John Schwarz likes to refer to this cancellation as the first miracle of superstring theory.

COMING FROM LEFT FIELD

To understand what all the excitement about superstring theory is, we have to go back to all those difficulties bedeviling the theory of gravity. For a good part of this century, many physicists have tried to unify gravity with the other forces, to marry gravity to the quantum, and to sneak a peek into the era of dark ignorance, but they all failed. As is often the case in physics, progress did not come from those banging their heads against the wall and moaning about gravity, but from an entirely different crowd. The string and the superstring theories were originally invented to describe the behavior of the strongly interacting particles of the subnuclear world—and failed miserably. It was not until 1974, when John Schwarz and the late French physicist Joel Scherk suggested that the theory was being applied to the wrong set of phenomena, that it should be used to describe gravity instead. And now, in 1984, Green and Schwarz were saying that the superstring theory, in those versions certified to be free from anomaly, is in fact a sensible theory of gravity.

STRING AND SUPERSTRING

Over the years, the language of physics has been refined steadily, arriving in our times at the enormous sophistication of the quantum field theory. Sophisticated though it is, the quantum field theory is ultimately still based on the simple intuitive notion that material objects can be divided into particles and that these particles are like tiny balls that can be treated mathematically as points, a notion that has been with physics almost from the beginning and that most physicists and I personally find rather appealing.

It is in fact astonishing that physics has managed to describe Nature so well for so long with point particles. When Democritus talked about atoms, when Newton talked about corpuscles, they did not insist on atoms and corpuscles being strict mathematical points, as far as I know. After the discovery of atoms, it soon became clear that atoms were not points but were made of smaller entities. But thus far, all the evidence indicates that the particles fundamental physicists deal with—quarks, electrons, photons, and the like—are mathematical points within the accuracy of the experiments. For instance, experimenters have determined that the electron can't be bigger than about 10^{-18} centimeter.

In contrast, the string theory, which begat the superstring theory, is based on the notion that fundamental particles are actually tiny bits of strings. Indeed, if the bit of string is much shorter than the resolution of our detection instruments, it will look like a mathematical point. What we thought were point particles flying around in space and time are actually itsy-bitsy pieces of strings flying around, according to string theory enthusiasts.

As the bits of string move around, they wriggle like tiny worms. The remarkable feature of the string theory is that by wriggling in different ways, the string can appear to us as different particles. In fact, the string theory contains infinitely many different particles, since a piece of string can wriggle in infinitely many different ways.

UNIFICATION OF GRAVITY

How does the string theory unify gravity with the other forces?

Recall that Einstein's theory of gravity specifies completely the properties of the graviton, the fundamental particle of gravity. The graviton is massless, spins twice as fast as the photon, and so on. Remarkably, the converse also is true: If a theory contains a particle that has precisely the same properties as the graviton, then that theory contains Einstein's theory.

As it happens, one of the infinitely many particles contained in the string theory has precisely the properties of the graviton. Thus the theory automatically contains Einstein's theory of gravity. The fact that the theory contains the graviton is hailed by the faithful as the second miracle of the superstring theory.

Gravity is unified with the electromagnetic interaction because another one of the infinitely many particles has precisely the properties of the photon. A theorem analogous to the one I cited above states that in this case the theory contains the electromagnetic interaction. Similarly, it turns out that the string theory contains the strong and the weak interactions as well.

The theory does not so much unify gravity with the other three forces as contain gravity and the other three forces.

Gee, you may say, that's awfully simple. I could have thought of it. Indeed you might have. The basic conceptual idea is almost embarrassingly simple. In contrast, the mathematical formulation is horrendously complicated. Part of this is due to our lack of familiarity with strings. For centuries, physicists played with point particles. To deal with strings, we suddenly have to extend our entire formulation of physics. What is worse, the intuition we have built by playing with point particles is by and large irrelevant.

What is so horrendously complicated about vibrating strings? Even introductory books on physics often talk about vibrating violin strings. Well, first of all, these strings are vibrating at the speed of light. Second, one has to keep track of the infinitely many ways in which the string can vibrate. Third, when one quantizes the theory, as one must to describe our quantum world, particles with totally unacceptable properties—particles that physicists have playfully dubbed ghost particles—threaten to appear. It takes an exquisitely elaborate analysis to show the ghost particles to be merely mathematical fictions. (To be fair, I must mention that when the quantum field theory was first invented, it also struck physicists as horribly complicated. With the passage of time, the horror gradually subsided—partly because physicists became familiar with the quantum field theory, and partly because they developed new ways of looking at the quantum field theory that made it seem a lot simpler. Many physicists believe that the same developments will happen to the string theory.)

STRINGS GO SUPER

Unfortunately, the basic theory of strings as I have described it, the sort of theory you and your Uncle John might have written down, turns out to be glaringly inadequate: It does not even contain the electron! That is because you and your Uncle John, being perfectly reasonable and sensible people, would have described the string vibrations in a reasonable and sensible way, something like the following: As one-tenth of a second has elapsed, this point on the string has moved west by 0.27 inch, this other point has moved west by northwest by 0.18 inch, and so on until the movement of every point on the string has been specified. You would have described how the string has moved from one particular instant in time to an instant one-tenth of a second later. To include the electron, physicists had to extend this description. They say that in one-tenth of a second this point on the string has moved in some direction by 0.27 inch and ψ, where ψ represents a weird sort of number invented by a mathematician named Grassman. Physicists represent Grassman numbers by a Greek symbol such as ψ.

What is a Grassman number? In a nutshell, a Grassman number

is a number such that when you multiply any Grassman number by itself, you get zero. Thus, $\psi \times \psi = 0$. Whaaat? Grassman numbers are weird. How can you possibly multiply a number by itself and get zero? Only a mathematician would think of such a thing.

A string theory in which the strings are allowed to move by amounts described by Grassman numbers as well as by ordinary numbers is known as a superstring theory.

I can't possibly give you in less than thirty pages a detailed explanation of how physicists use Grassman numbers, and I am sure you would be bored stiff. I mention Grassman numbers only to show you one weird aspect of the theory. Once the theory is detached from its moorings on ordinary numbers, physicists have a hard time visualizing it as describing vibrating strings. If you think string theory is abstruse, superstring theory is even more so.

Not impressed! you say. What is so profound about going from points to strings? Nothing, physicists would have to concede if hard-pressed. Then why did it take so long for physicists to move from a theory of points to a theory of strings? The answer is that physicists had been making such tremendous progress with theories based on point particles that there was not a strong motivation to go to theories based on strings. Actually, over the years there had been sporadic attempts to write down theories not based on point particles, but physicists had generally recoiled in horror at the attendant complications. One must pay tribute to pioneers of the string theory such as John Schwarz and Michael Green for their persistence.

BEYOND STRINGS?

Heck, I can make a new theory, too, you exclaim. If you guys just moved on from theories of points to theories of strings or curved lines, why stop? As mathematical objects, points are zero-dimensional, lines are one-dimensional, surfaces are two-dimensional, and ball-like blobs are three-dimensional. Why not continue the natural progression and move on to theories of surfaces and then theories of blobs? Just as a teeny bit of string can look like a point particle, so a teeny bit of surface or membrane can look like a teeny bit of string. Why not a membrane theory and a supermembrane theory, a blob theory and a superblob theory?

People have tried to write down membrane theories and blob theories, but these theories are so complicated that physicists couldn't get past square one in analyzing them. But surely, Whoever designed the universe couldn't care less whether humans in the late twentieth century should have enough mathematical pizzazz to solve the puzzle. Of course, the string theory is presumably as good an approximation of the membrane theory as the point particle theory is an approximation of the string theory. String theory enthusiasts would concede that theirs might not be the ultimate theory either but merely a better theory than the point particle theory. Still, many physicists, and I for one, are bothered that by moving away from points, physicists might be opening up an infinite regression of Russian dolls.

AN EXTRAVAGANT UNIFICATION

Gravity is finally unified with the three other fundamental forces, in the sense that the superstring theory contains the graviton and the fundamental particles associated with the other forces, such as the photon. But the theory also contains an infinite number of other particles. These particles all have enormous masses, at least 10^{19} times the mass of the proton. According to the terminology introduced in the preceding chapter, these particles have masses of the order of the Planck energy and beyond. At "ordinary" energies—that is, at energies far below the Planck energy—they are too massive to participate. They come in only when the energy available exceeds their mass.

One might say that gravity is unified with the strong, the electromagnetic, and the weak forces, but at the same time also with an infinity of other forces that do not show up at ordinary energies. It is what might be called an extravagant unification. Gravity has been persuaded to join the dance, but only at the cost of hiring an infinite number of extras.

How does string theory manage to marry gravity to the quantum? Recall that when physicists tried to turn Einstein's theory of gravity into a quantum theory, the quantum fluctuations with energy above the Planck energy produce an infinitely large correction to the gravitational force between two objects. How does the string theory get around this difficulty? Roughly speaking, when the energy of the quantum fluctuations exceeds the Planck scale, all those "extras" in the drama of physics, the whole infinitude of them, come to the fore and sing their song. Hey hey, they sing, at such high energy, we can dance to the tune of the quantum and generate quantum fluctuations like everybody else. These fluctuations also produce an infinitely large correction to the gravitational force, which, miracle of miracles, cancels the unacceptable correction produced in Einstein's theory.

To put it in another way, Einstein's theory of gravity is merely an approximation of a piece of the string theory for energies below the Planck mass. For energies above the Planck scale, it's another ball game, and calculations of quantum fluctuations using Einstein's theory do not tell the whole story.

MULTIDIMENSIONAL UNIVERSE

An astonishing feature of the superstring theory is that it can be formulated only if space is nine-dimensional. Upon first hearing this, you might want to chuck the theory out the window, and that was indeed the initial reaction of many physicists. Space is evidently three-dimensional to any sensible person. Einstein had unified space with time and described the physical world as four-dimensional: the three familiar dimensions of space and the dimension of time. But to Einstein, space remains three-dimensional, as any ordinary person would have thought. The superstring theory, however, makes sense only if there are six extra dimensions to space.

How could we possibly have missed these extra dimensions? Easily, if these extra dimensions are tiny. Consider a creature constrained to live on the surface of a tube. The space inhabited by the creature, the surface of the tube, is really two-dimensional. Suppose the radius of the tube is much smaller than the smallest distance that can be perceived by the creature. To the creature, space would appear to be one-dimensional; since the creature can move only along the tube.

In other words, we can mistake a very thin tube for a line. On closer inspection, every "point" on the "line" actually turns out to be a circle. Thus it is quite possible that every point in the familiar three-dimensional space in which we move would, on closer inspection, also turn out to be a circle. If the radii of the circles are much smaller than the smallest distance we can measure, we would see the circles as points, and we would be misled into thinking that we are living in a three-dimensional space instead of a four-dimensional space. Don't worry if you can't visualize such a strange space. Neither can the myopic creature living on the long, hollow tube visualize that his space actually is two-dimensional.

We can keep going. Perhaps on closer inspection, each point in our three-dimensional space would turn out to be a two-dimensional surface, such as the surface of a sphere. In that case, our beloved three-dimensional space would actually be five-dimensional, and so on. Physicists call the three-dimensional space that we know external space and the space hidden from view because of our myopia internal space. The actual dimensions of space are then the dimensions of internal space plus three.

Experimentally, there is no evidence whatsoever of an internal space. Experimenters have looked and said that the internal space, if it exists at all, must be a few hundred times or so smaller than the size of the proton. Now, that is absolutely teeny compared to the human scale of things, and we must congratulate the experimenters for their heroic efforts in looking down to such tiny scales. But while experimenters work hard, theorists are free to believe in an internal space. They merely have to snap their fingers and chant in unison that the internal space is much smaller than what the experimenters can measure. In fact, string theorists believe the size of the internal space to be 10^{18} times smaller than the size of the proton. There is no hope for experiments to be able to detect a space of this size within the foreseeable future.

ELECTROMAGNETISM OUT OF GRAVITY

The idea that space has hidden internal dimensions is in fact recovered from the dustbins of history. In 1919, just four years after Einstein proposed his theory of gravity, the Polish mathematician and linguist Theodor Kaluza suggested that space may be four-dimensional. The Swedish physicist Oskar Klein then developed Kaluza's work into what is known as the Kaluza-Klein theory. By the way, when Einstein learned of these ideas, he was astonished. He wrote to Kaluza that the notion of four-dimensional space had never occurred to him.

Any sensible person might react to Kaluza's suggestion by asking what good an extra hidden dimension of space is. You can assert that any sort of fancy chimera exists just by insisting that it is too small to be seen.

To be sure, for physicists to have paid any attention at all to Kaluza and Klein, they must have found a wonderful benefit in hidden dimensions. They discovered that the Kaluza-Klein theory unifies gravity with electromagnetism.

Consider the simplest possibility: that the internal space is a circle and that the world is actually five-dimensional, with four dimensions for space and one dimension for time. Suppose, Kaluza and Klein said, that there is only gravity in the world. How would the inhabitants of this world, too myopic to see that what they call points are actually circles, perceive the gravitational force? To Kaluza and Klein's utter surprise, they found that the inhabitants would feel two types of forces, which they could interpret as a gravitational force and an electromagnetic force! In Kaluza and Klein's theory, electromagnetism comes out of gravity!

We can understand this stunning discovery roughly as follows. A force in a three-dimensional space can pull in three different directions; after all, that is what we mean by saying that space is three-dimensional. In the four-dimensional space of the Kaluza-Klein theory, gravity can pull in four different directions. To us, the myopic inhabitants, a gravitational pull in the three directions corresponding to the three directions we know appears as just a gravitational pull. But what about a gravitational pull in the fourth direction, the direction that we are too myopic to see? We would construe that as another force and call it electromagnetic.

To many physicists, the emergence of electromagnetism from gravity can only be described as "mind-blowing." Nevertheless, the Kaluza-Klein theory faded away from the collective consciousness of physicists as it became clear that there was more to the world than gravity and electromagnetism. The strong and the weak forces had to be invented to account for a whole host of new phenomena, and the strong and the weak forces, as they were understood, just did not fit into the Kaluza-Klein framework. When I studied physics, I never heard of the Kaluza-Klein theory. The major textbooks on gravity in the 1970s never mentioned it.

FORCES FROM GEOMETRY

Even while the Kaluza-Klein theory was in disrepute, a few physicists went back to the theory periodically. Kaluza and Klein took the internal space to be a one-dimensional circle. What if the internal space is multidimensional? For example, the internal space could be a two-dimensional sphere. (See Figure A.1a.)

These physicists found that the more dimensions the internal space has, the larger the variety of forces that emerge. If the internal space is a circle, the electromagnetic force emerges. If the internal space is a sphere, a collection of forces containing the electromagnetic force emerges. This is perhaps not surprising in light of the argument given above: There are more directions on the sphere in which forces can pull than on the circle.

How do these forces depend on the symmetry of the internal space? For example, a two-dimensional internal space could be a sphere or a torus. The torus is symmetric under rotations around the axis indicated by the dotted line in Figure A.1b, while the sphere is symmetric under rotations around any axis. It turns out that if the internal space is a sphere, more symmetries relate the forces that emerge than would be the case if the internal space is a torus. Indeed, if the internal space has no symmetry at all, as indicated by the space in Figure A.1c, then no force emerges. The number of forces that emerge depends on how symmetric the internal space is. Thus, more forces emerge from the sphere, which is more symmetrical than the circle.

"It is so incredibly neat!" a Kaluza-Klein aficionado would exclaim. The geometrical symmetry of the internal space imprints itself on the symmetry of the forces we see in the external world.

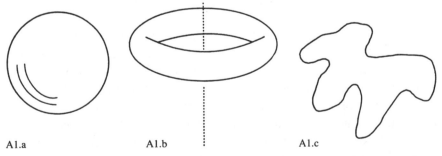

A1.a   A1.b   A1.c

Possible choices for the internal space. Figure A.1a. A sphere. Figure A.1b. A torus. Figure A.1c. A surface of no particular symmetry. For the meaning of the dotted line in A.1b, see the text. In the figure, the space is by necessity represented as two-dimensional surfaces. In fact, physicists typically think of higher-dimensional spaces.

Physicists have gone into ecstasies over the symmetries of the fundamental forces. Meanwhile, mathematicians have long exulted over the symmetries of geometrical shapes. In the Kaluza-Klein theory, the symmetries of the mathematicians come together with the symmetries of the physicists. This metamorphosis of geometrical symmetry into physical symmetry is extremely beautiful to watch, but unfortunately it can be appreciated fully only when cloaked in the splendor of mathematics.

By the 1970s, physicists had reached a new understanding of the strong and the weak forces. With this new understanding they managed to unify the strong, the electromagnetic, and the weak forces into one. It was then realized that this grand unified force could emerge from the Kaluza-Klein theory as a piece of gravity. At that point, the Kaluza-Klein theory came roaring back. Physicists tried different internal spaces. I for one was convinced that the internal space has to be a sphere, that most aesthetically satisfying of geometrical shapes.

I should emphasize that the idea of space having more than three dimensions is by no means universally accepted. Why is the internal space so small while the external space—namely, the whole universe—is so huge? Different dimensions of space presumably started out on the same footing, but for some unaccountable reason, some of them have stretched out, while others have shriveled up to almost nothing. Some physicists consider the whole idea of hidden unseen dimensions as simply too farfetched and too removed from experimental reality.

A BROKEN PROMISE

At our present stage of understanding in physics, the masses of fundamental particles such as the electron are simply regarded as undetermined parameters whose values are known merely because they are measured. Why is the electron's mass equal to about half an MeV? Nobody knows. The Kaluza-Klein theory, however, asserts that the masses of fundamental particles are determined by the size of the internal space.

Physicists were excited by this bold promise of the Kaluza-Klein theory. Unfortunately, a quick calculation immediately threw cold water over their excitement: The electron's mass came out to be about the Planck mass, 10^{19} GeV. The theory is a dog! The answer, 10^{19} GeV—

that is, 10^{21} MeV—is not even in the same ballpark as half an MeV. In the Kaluza-Klein theory as we know it, the smaller the internal space, the larger the masses of particles like the electron. Since we can't make the internal space large, we can't make the electron mass small.

In the late 1970s and early 1980s, physicists tried a variety of ways of modifying Kaluza-Klein so the electron would come out to have a small mass. Nothing worked. In frustration, they were ready to give up on the theory.

It was more or less at this point that the superstring theory came on the scene. As we were saying before we launched into this long stretch of background on the Kaluza-Klein theory, the superstring theory makes sense only if space is nine-dimensional. With all this ongoing flirtation with the Kaluza-Klein theory, physicists took the news calmly. Far from chucking the theory out the window, they immediately absorbed the Kaluza-Klein idea into the superstring theory.

In what is sometimes known as the third miracle of the superstring theory, the electron comes out to be massless. Now, you might not think that this should count as a miracle, since the electron did not come out to have a mass of half an MeV. Well, compared to the Planck mass scale, half an MeV is for all intents and purposes equal to zero. Certainly it is a huge improvement over the Kaluza-Klein theory. Believers in the superstring theory are hoping that eventually, when the theory is well understood, some small effect now overlooked will correct this predicted value of the electron's mass from zero MeV to half an MeV. At this stage of the game, it is regarded as a minor detail.

TOO MUCH OF A GOOD THING

The Kaluza-Klein idea, however, could be implemented in the superstring theory only in a rather strange way. The difficulty is an embarrassment of riches. I mentioned that the string wriggling in different ways appears as different particles. Wriggling in one way, the string appears like a graviton; wriggling in another, it appears like a photon. Thus the theory contains the photon and the associated electromagnetic force. Similarly, it contains the strong and the weak forces. The theory already contains these known forces; it needs the forces that would come out of a Kaluza-Klein framework like it needs a hole in the head.

The string theorists are really in a bind. Physicists have striven to understand where the four known forces come from. Now suddenly there are too many forces: the forces from the wriggling string and the forces from the Kaluza-Klein framework. To avoid getting these Kaluza-Klein forces, string theorists are forced to choose internal spaces with no symmetry.

As I mentioned earlier, if the internal space is completely nonsymmetrical, forces do not emerge. By choosing a nonsymmetrical internal space, string theorists manage not to have too many forces: They keep only those from the wriggling string.

Traditionally, Kaluza-Klein aficionados have always considered symmetrical internal spaces. Why would anyone consider nonsymmetrical internal space? After all, the whole point of the Kaluza-Klein theory is to put in gravity and get out some other force besides gravity. In this sense, the implementation of the Kaluza-Klein idea in the superstring theory is against the original spirit of Kaluza and Klein. Some physicists, myself among them, are bothered by this subversion of the Kaluza-Klein theory. The beauty and advantage of going to a higher-dimensional space is that other forces emerge as pieces of gravity. But the superstring theory does not need these forces and so uses (according to some beholders, at least) the yuckiest possible internal space.

I for one am still hoping that the internal space will turn out to be a perfect sphere and that the original beauty of the Kaluza-Klein theory will be preserved. But my superstring friends dismiss this sort of sentiment as wishful thinking and ill-becoming nostalgia.

The complications of the superstring theory are due partly to the symmetrical internal space. With a nonsymmetrical internal space, physicists have great difficulties working out the consequences of the theory.

In Chapter 9, I mentioned how topological notions have entered physics and how by means of topology mathematicians can deduce properties about geometrical objects ordinary mortals cannot even visualize. In the Kaluza-Klein theory and the superstring theory, topological considerations really came to the fore. As long as the internal space is symmetrical like a sphere, its properties can be deduced readily. But when the internal space can barely be visualized, as in the superstring theory, then topological methods are needed to extract the answers to such physical questions as whether the electron's mass can be small compared to the Planck mass.

In the initial excitement, it was thought that the superstring theory could determine the internal space. Alas, that promise has proved to be illusory thus far. Many different types of internal spaces are possible, and to each of these spaces corresponds a different superstring theory. Physicists now have to rely on topology and other branches of higher mathematics such as algebraic geometry to help them classify and study these spaces.

HERE LIE DRAGONS

My lady friend is grumbling, "I still don't understand the superstring theory."

"Well, I know," I say with a sigh, "I am sorry. I didn't explain the superstring theory so much as describe it. Partly that's because the theory is both physically and mathematically so novel that physicists don't have a good understanding of what's going on. I didn't explain why the theory is not afflicted with the anomaly or why the electron mass comes out to zero. Nobody really understands all these *whys*. These miracles, and that's why they are jokingly referred to as miracles, just appeared out of the equations."

"That made me feel better. I was worried that I may have missed something." She appeared relieved.

"It doesn't make me feel any better. In writing a book on physics for an intelligent lay reader like you, my ideal is to try to explain as much as possible. But when I come to the latest, so much of it is based on the subtleties of quantum physics," I said rather ruefully.

"Well, that figures. I can at least picture a string wriggling. By wriggling in different ways, it can appear to us as different particles," she replied cheerfully.

"Right, that's the key point. The graviton is just one of an infinitude of wriggling forms. In a way, gravity has been demoted. That's one reason why some physicists are unhappy with the superstring theory. Somehow we feel that gravity should be special, being connected to space and time and all."

"Hmm, that kind of bothers me, too."

"But no doubt about it, superstring theory is by far the most significant development in our understanding of gravity since 1915. Ever since the arrival of quantum physics, gravity has pointed to its own demise. You know how mapmakers used to mark those regions they don't know anything about with a Latin inscription proclaiming, 'Beyond here lie dragons.' Gravity has marked the energy domain beyond the Planck energy with that inscription. Superstring theorists have now ventured into that region. Well, there are dragons! They are the wriggling string in its infinite number of forms."

MIRACLES AND DIFFICULTIES

With three miracles, the faith of the superstring enthusiasts was sealed. However, there are also difficulties and complications. At the moment, the physics community is divided over whether the superstring theory will turn out to be the ultimate theory of everything, as the enthusiasts proclaim. The theory is enormously complicated to develop, and the stage has not yet been reached when the issue can be settled decisively by experiments. An intrinsic difficulty is that while experiments can be conducted only at energies far below the Planck energy, the superstring theory is formulated expressly at that lofty energy scale. All those infinite numbers of extras introduced by the superstring theory—the supermassive particles representing all the different ways in which the string can wriggle—come out to sing their song only at the Planck energy and beyond.

RUNGS ON A LADDER

To understand better the differences in opinion over the superstring theory, let us discuss the energy scales of physics in more detail. These energy scales may be visualized as rungs on a ladder. On each rung physicists are trying to work out all the phenomena characterized by that energy scale. Consider, for example, an atomic physicist trying to calculate the properties of the sodium atom, which happens to have eleven electrons whizzing around a nucleus. As far as he is concerned, the sodium nucleus is just a tiny ball characterized by a few numbers such as its mass, charge, and magnetic moment. (A nucleus's magnetic moment tells us how it would behave in a magnetic field, just as its charge tells us how it would behave in an electric field.) He regards these numbers as givens, either directly measured or calculated, handed to him by a nuclear physicist. The nuclear physicist, on the other hand, tries to calculate the properties of the sodium nucleus knowing that it is made of eleven protons and twelve neutrons. She can't do her job, however, unless she is told about the properties of the proton and the neutron. Okay, if you tell me that the magnetic moments of the proton and of the neutron are such and such, then I can calculate the magnetic moment of the sodium nucleus. The particle physicist, in his turn, tries to calculate the magnetic moments of the proton and of the neutron.

Thus there is an orderly progression in our understanding of the physical world. On any given rung, physicists have to be given a bunch of numbers, sometimes known as parameters. As we move up the rungs of the ladder, we hope that the number of parameters needed at each rung diminishes. Thus, atomic physicists treat the magnetic moments of some several hundred-odd atomic nuclei all as parameters, while nuclear physicists regard the magnetic moments of only the proton and the neutron as parameters. (As far as magnetic moments are concerned, the story actually ends at the particle physics rung. The presently accepted particle theory fixes the magnetic moments of quarks with some definite values, while the magnetic moments of the proton and the neutron can be calculated in terms of the magnetic moments of the quarks.)

Even as physics explores ever higher energy scales, a myriad of phenomena remain to be understood at lower energy scales. For instance, the behaviors of various metallic substances continue to confound the expectations of physicists. However, the explanations of these phenomena surely do not lie in the physics of higher energy scales. The disintegration of a subnuclear particle in a zillionth of a second is hardly going to affect the behavior of metals.

Our understanding of the physical world thus can be quantified roughly by giving the number of parameters needed. The currently accepted particle theory contains eighteen parameters. (The precise number depends on how you count and what you regard as understood.) The goal of fundamental physics is to reduce the number of parameters down to an absolute minimum. Some physicists even entertain the hope that ultimately the number of parameters will be reduced to zero.

Where is the next rung on the ladder after particle physics?

Physicists can only guess. From atomic physics, with its characteristic energy scale of 10 eV, we had to climb up a hundred thousand times higher in energy to reach nuclear physics, with its characteristic energy scale of 1 MeV (that is, one million electron volts). From nuclear physics we had to climb up a thousand times in energy to reach particle physics, with its characteristic energy scale of 1 GeV (that is, one billion electron volts). At present, some aspects of physics up to an energy scale of 100 GeV have already been explored at high-energy accelerators, but nothing dramatically new has been seen. (I am using "new" in a strict sense. To qualify as new, the physics involved has to be such that it requires structural modifications to the presently verified theory.)

Will we see new physics at 1,000 GeV? Or only at a much higher energy scale?

Much rides on the question of where the next rung is. Lots of taxpayers' money is involved. The physicists who are building the next generation of high-energy accelerators would dearly like to know. Should they design the machine to explore physics at 1,000 GeV (known as 1 TeV), or should they shoot for 100 TeV?

All we know for sure is that there is a rung at the Planck scale, at 10^{19} GeV. Gravity tells us so. Furthermore, the grand unified theory suggests that the strong, electromagnetic, and weak interactions merge into one interaction at 10^{15} GeV, but not everyone believes in grand unification. Meanwhile, we have explored physics experimentally only at energy scales up to about 10^2 GeV.

An extreme view holds that that's it, boys and girls. There's not gonna be any dramatically new physics between 10^2 GeV and the grand unification scale. Grand unification enthusiasts call this enormous region "the desert," in which they say, experimenters will find nothing but "barren sand." Experimenters and physicists involved in building new accelerators hate this view, naturally. Outraged, they point to the history of physics. In the past, whenever we went up in energy, we encountered dramatically new physics whose existence we had had no inkling of. Surely, "the desert" will bloom and prove to be a land of milk and honey.

PHILOSOPHICAL DIFFERENCE

Viewed against this background, the difference in opinion over superstring theory reflects a real philosophical difference among physicists on how physics should progress: by great leaps upward, or by arduously climbing up rungs of the ladder. The superstring believers hold that the true physics lies at the Planck scale and that we have essentially understood, at least in broad outline, all of physics up to the Planck scale. In their boundless faith, they proclaim that they have found the Ultimate Design. In contrast, the infidels are people of little faith. They want their beliefs anchored in hard facts, not in the soft elegance of higher mathematics. They fear that their colleagues, in their joyous leap toward the Planck scale, may instead have arrived at a never-never land of mathematical enchantment.

This clash in philosophy has always existed, between the idealists and the empiricists, between Einstein and the vast majority of theoretical physicists whose days are largely spent pondering over data and performing routine calculations.

Einstein once said, "I want to know how God created this world. I am not interested in this or that phenomenon. I want to know His thoughts; the rest are details." What arrogant idealism! What provocation! Hear the gnashing sound of anger coming from those physicists who are dedicating their lives to this or that phenomenon!

Of course, most of physics, from the study of stars to the study of metals and materials, is concerned precisely with this or that phenomenon, and rightly so. We are speaking here of what may be called fundamental physics, the striving to understand how the world is put together ultimately, and of the philosophical differences that exist within the fundamental physics community itself.

The old man is unrepentant and speaks again: "I am convinced that we can discover by means of purely mathematical construction the concepts and the laws . . . which furnish the key to the understanding of natural phenomena. Experience may suggest the appropriate mathematical concepts, but they most certainly cannot be deduced from it. . . . In a certain sense, therefore, I hold it true that pure thought can grasp reality, as the ancients dreamed." Pure thought! What hubris, but what glorious hubris! This is the dream that lures boys and girls into becoming real theoretical physicists! Whew, understanding the world through pure thought!

The skeptics are snickering. Yeah, the ancients dreamed, but they didn't get very far. Democritus thought matter was made of atoms, and in cultures from Islamic to Chinese there were sages who had similar ideas. But we would have known nothing about atoms without the experiments of the late nineteenth and early twentieth centuries. Mao felt that dialectical materialism (remember Gamow and his exam!) required subnuclear particles to be made of "stratons" (from the word "stratum"). But without the big accelerators built in the 1950s and 1960s we probably would never have found out about quarks and gluons.

What then was Einstein talking about? Did physics ever progress by pure thought? Well, the best example (and, some would argue, the only example, if it's an example at all) is provided by the old man's own theory of gravity. From the thought that a falling man feels no gravity flow the secrets of gravity. Of course, the thought itself had to be built on a fact, in this case Galileo's observation that all objects fall at the same rate. The point is that the theory was not constructed laboriously step by step, with experiments providing guidance at each turn. It was born whole. Never mind that the birth process was painful—Einstein had to struggle for years—the theory was born whole.

But Einstein's theory is the exception to the rule. The struggle to understand the weak interaction, for example, is more closely characteristic of the development of fundamental physics. From the discovery of radioactivity to the present, the theory of weak interaction was built brick by brick, with bricks often removed and even whole wings demolished when new experiments contradicted the theory.

Einstein's sentiment reflects his intoxication with his own success in discovering the theory of gravity. Many felt, however, that the discovery of the theory of gravity was the one exceptional case in which pure thought could take physics far. Three quarters of a century later, believers in the superstring theory are once more flying the banner of pure thought. The notion that fundamental particles are in reality wriggling strings was a pure thought, unmotivated by any experiment.

Particle physicists are deeply divided. While some revel in "pure thinking," others feel that Einstein's dream—if it is represented by the superstring theory—has turned out to be a nightmare. In practice, the decision of an individual physicist to work or not to work on the superstring theory is often based less on a grand philosophical outlook than on other factors, such as temperament, ability, prospects for career advancement and, perhaps most importantly, sloth and inertia. It takes a great deal of energy to master a radically new development, and many physicists, even if only half skeptical of superstring theory, have opted to continue doing whatever they have been doing.

Notes

Some of the following notes are bibliographical references, others amplify or supplement discussions found in the text.

Chapter 1: In Search of Beauty

- Page 3. The passage quoted from Bondi can be found in *Einstein: The Man and His Achievement,* edited by G. J. Whitrow (New York: Dover, 1973).
- Page 3. Einstein's criterion in judging a physical theory is described by B. Hoffman in *Albert Einstein: Creator and Rebel* (New York: Viking Press, 1972). In describing Einstein's work, the author, a one-time collaborator of Einstein's, said, "The essence of Einstein's profundity lay in his simplicity, and the essence of his science lay in his artistry—his phenomenal sense of beauty."
- Page 4. Einstein expressed his wonder at the apparent fact that the world has a design, and that that design is comprehensible to us, in a letter to his friend Maurice Solovine. See "What, precisely, is 'thinking'? Einstein's answer," by G. Holton in *Einstein: A Centenary Volume,* edited by A. P. French (Cambridge, Mass.: Harvard University Press, 1979).
- Page 6. The distinction between phenomenological laws and fundamental is not completely clear-cut, of course. Einstein derived Newton's law of gravitation, once thought to be fundamental, as a phenomenological manifestation of his theory of gravity, but, recently, some theoretical physicists have demonstrated that Einstein's theory may, in turn, follow from a deeper theory. The hubris of physicists notwithstanding, what is regarded as fundamental by one generation may be regarded as phenomenological by a later one.

Chapter 2: Symmetry and Simplicity

- Page 9. A classic, but outdated, reference on symmetry in physics and mathematics is H. Weyl, *Symmetry* (Princeton, N.J.: Princeton University Press, 1952). See also E. P. Wigner's *Symmetries and Reflections* (Bloomington, Ind.: Indiana University Press, 1967).
- Page 17. The explanation of why stars burn slowly originated with Hans Bethe and others. For a good technical discussion, see D. D. Clayton, *Principles of Stellar Evolution and Nucleosynthesis* (New York: McGraw-Hill, 1968).
- Page 17. In the section "The Rule of Large Numbers," I talk about the mind-boggling number of photons and protons in the universe. The reader may wonder who counted all those photons and protons in the universe, never mind who ordered them. The census was totally unplanned. About twenty years ago, Arno Penzias and Robert Wilson, two engineers at the Bell Telephone Laboratories in Holmdel, New Jersey, built an ultrasensitive antenna. To their dismay, it produced a steady hum, despite their best efforts, which included, incidentally, periodically removing deposits left by some pigeons who took a liking to the antenna. It turned out that they were listening, in fact, to the song of the universe. The universe is suffused with microwave radiation, much like the space inside a microwave oven, but at an enormously lower intensity, of course. This great discovery, for which a Nobel prize was awarded, also helped to establish George Gamow's theory of the universe originating from a big bang. In detecting microwave radiation, Penzias and Wilson were actually seeing the faint glow of an explosion that occurred a long, long time ago. Microwave radiation, like radio and light waves, is a form of electromagnetic radiation. Knowing the number density of photons contained in microwave radiation of a given intensity from the theory of electromagnetism, and knowing roughly the size of the universe, the physicist has merely to multiply two numbers together to obtain the population of photons in the universe. Incidentally, the number of photons produced by all the stars in the universe since stars were born, let alone the number of photons produced by all our light bulbs, is minuscule compared with the number of photons contained in the cosmic microwave background.

    To determine the number of protons, one can simply count and multiply the number of protons in a typical star (such as the sun), the number of stars in a typical galaxy (such as our Milky Way), and the number of galaxies in the observable universe. But since the number density of photons is known quite accurately by measuring the cosmic microwave, it is better to determine the number of protons indirectly by first ascertaining the ratio of the number densities of protons to photons. In the early universe (early by human standards, but very late by the standards of particle physics), protons and neutrons were cooked into various nuclei. Most of the protons managed to marry an electron, thus forming hydrogen atoms; some, however, collided with neutrons and got stuck together to form a helium nucleus, for example. By tasting a cake, an expert cook could easily deduce the relative amount of flour to butter that went into the baking. In exactly the same way, the amount of helium that astronomers now observe in the sky tells us about the ratio of photons to protons. The imagery of cooking, incidentally, is more than picturesque in this instance: In baking, chemical reactions rearrange molecules. In the early universe, nuclear reactions bonded protons and neutrons.

Chapter 3: The Far Side of the Mirror

- Page 22. The quotation about passing directions at the dinner table comes from Judith Martin, *Miss Manner's Guide to Excruciatingly Correct Behavior* (New York: Warner Books, 1982), p. 130.
- Page 25. The description of the *Poeciliidae* family of fish comes from G. Murchie, *The Seven Mysteries of Life* (Boston: Houghton Mifflin, 1981), p. 134.
- Page 25. I would like to thank John Martin, one of my art history professors at Princeton, for a helpful conversation in which he told me how Rembrandt did not bother to adhere to the left-right convention in his etchings.
- Page 26. Men's striped ties provide another amusing example of human convention regarding left and right. In the United States, the convention that the stripes go from right shoulder to left hip has emerged over the years. In England, it's just the opposite. Some years ago, Brooks Brothers, a well-known U.S. clothier, decided to introduce a line of striped ties in which the direction of the stripes were reversed. The new pattern lasted only one season. That such a totally arbitrary convention should have such a powerful hold on us is indicated in a report from *Harvard Magazine* (1985).
- Page 26. In military weddings, such as those held at West Point, the bride stands to the right of the groom, so as not to be struck should he draw his sword.
- Page 28. I have drawn on the account of parity violation presented by J. Bernstein in *A Comprehensible World* (New York: Random House, 1967), p. 35. For another account of symmetry and parity violation, see C. N. Yang, *Elementary Particles* (Princeton, N.J.: Princeton University Press, 1961).
- Page 33. For a brief biographic sketch of Madame Wu, see G. Lubkin, "Chien-Shiung Wu, the First Lady of Physics Research," in *Smithsonian,* January 1971. For a discussion of the history of experiments on the

weak interaction, see C. S. Wu, "Subtleties and Surprise: The Contribution of Beta Decay to an Understanding of the Weak Interactions," in *Five Decades of Weak Interactions,* Annals of the New York Academy of Sciences, Vol. 294, 1977.

- Page 33. Sure that Madame Wu would have a fascinating story to tell, I traveled to Columbia University to chat with her, having taken care to obtain a parking permit in advance. An energetic, pretty lady, with a prosperously patrician air about her, Madame Wu belies completely whatever stereotypic image one may have of a leading experimental nuclear physicist. Moving her delicate hands in excitement, she chuckled as she reminisced about the personalities and events in her life.

Words came easily. She recalled that, as a young girl, she viewed with awe the sword and gun that her father had used in the revolution of 1911. During the Second World War, she worked on the Manhattan project ("Right here in Manhattan," she said with a laugh, "in a laboratory hidden in an auto showroom between 136th and 137th streets"). When I asked her about her famous experiment in which she first saw the intrinsic difference between the mirror world and our world, her eyes lit up and she said, with evident pleasure, "That was a lot of fun!" But how did a girl born almost during the Manchu dynasty into a feudal and male-dominated society come to be known as the "reigning queen of experimental nuclear physics," and become the first woman president of the American Physical Society?

Madame Wu explained that she was born in Liu Ho, situated on the mouth of the Yangtze River, a town which, by virtue of its location, figured in history as the launching point of Imperial naval expeditions, and, later on, as one of the first places in China exposed to Western influence. Because of silting, the town eventually lost its importance as a port to another small town about twenty miles upstream, Shanghai. Madame Wu's father had gone to Shanghai to clerk in one of the foreign trading companies. He was so imbued with modern notions that after some years he decided to return to his hometown to start a girls' school, much to the displeasure of his father. Madame Wu recalled that her father had learned to build radios in Shanghai. Starting around 1930, broadcasts from Shanghai could be received in Liu Ho, and so her father would construct radios for the peasants.

But her childhood was far from totally tranquil. Thanks to coastal pirates and warlords, Liu Ho was ravaged repeatedly. She was luckier than most, however, and completed her university studies in 1936 at the National Central University in Nanjing, a university founded surprisingly, perhaps, in 1902. After working briefly on X-ray research at Zhejiang University, under a woman who had studied at the University of Michigan, Madame Wu had the opportunity to study further at Michigan. However, on arriving in San Francisco she was taken by a friend to see Berkeley; she liked the campus at once. Furthermore, Madame Wu recalled that she had very much wanted to immerse herself in American life, and she learned that, for some reason, most of the students from China went to Michigan. She decided not to go. A physics student she had met took her to see Raymond Birge, a physicist now generally credited with building up the physics department at Berkeley. Madame Wu recalled that Birge was "very, very nice," and he allowed her to enroll even though classes had already been under way for several weeks.

In 1939, word came to the United States that two German physicists, Otto Hahn and Fritz Strassmann, had managed to split the uranium nucleus, thereby causing a large amount of energy to be released. Ernest Lawrence, the great South Dakotan who invented and developed the particle accelerators, without which modern nuclear and particle physics would be impossible, put Madame Wu to work immediately. She was given clearance, and the expertise she gained was later to be crucial in the Manhattan project when one of the atomic piles built by Fermi stopped working after a few hours. In 1942, Madame Wu took up a teaching position at Smith College, the well-known women's institution located in pastoral Massachusetts. That might have been the end of her research career had not Ernest Lawrence come to visit. He remarked that she really ought to be doing research. Soon afterward, she was offered research positions at Harvard, Columbia, Princeton, M.I.T., and several other research centers, such was the influence of Lawrence. She went to Princeton and found that "the boys were very nice," particularly the nuclear physicist Henry Smythe, who tried very hard to place her on the all-male faculty. (Later, she was to be the first woman awarded an honorary degree by Princeton.) When Columbia asked her again the next year, Madame Wu decided to move to New York.

In her position, Madame Wu is often asked about women going into physics. She is very pleased that women nowadays are confident that they can do anything. Recalling that women physicists already have had a major impact, Madame Wu exclaimed, "Never before have so few contributed so much under such trying circumstances!" She feels lucky that in her career everyone has been very nice to her. Contrary to what one might have guessed, she met no resistance in China. There was an enormous sense among her fellow university students that since China was so backward, both sexes must work hard in all fields. After her arrival in the United States, she learned to her amazement that women were excluded from most of the leading private universities. She was even more astonished when, many years later, one of her women graduate students told her that boys do not like to go out with girls who are studying physics. At Berkeley, Madame Wu was the only woman in physics. Every one was "most helpful and kind." Robert Oppenheimer, who taught her quantum mechanics, was a "perfect gentleman."

Madame Wu is saddened that other women physicists were not all so lucky. Lise Meitner (1878–1968), an Austrian physicist working in Germany and a pioneer researcher of radioactive decays, was forced to do her experiments in a carpenter's shed outside the physics building. Meitner was the first person to understand nuclear fission. When she visited Columbia, Madame Wu, mindful that older people have to go to the bathroom more often, asked Meitner periodically if she would like to go. Meitner had replied wryly, no, she had been well trained. There was no ladies' room in the carpenter's shed.

- Page 35. Much of Ellis's work was done with W. A. Wooster.
- Page 37. The poem "Cosmic Gall" can be found in John Updike's *Telephone Poles and Other Poems* (New York: Knopf, 1965).
- Page 37. As an illustration of how yesterday's physics could become tomorrow's technology, it has been suggested that the ghostlike neutrino could be used to prospect for oil. The idea is to send a beam of neutrinos through the round earth and to study it emerging some distance away. Since oil and rock have different nuclear compositions, the number of neutrinos that get through will tell us, in principle, whether or not the beam has passed through an oil deposit. Given the present difficulty in producing and in detecting neutrinos, the scheme is rather futuristic. But who knows, one day, neutrinos may enable us to peek into places now off-limits. (Incidentally, beams of another particle, the muon, have already been successfully used to detect secret chambers in the pyramids.)
- Page 37. For half a century, physicists have repeated Ellis's measurements with ever-increasing accuracy, in an attempt to determine whether or not the neutrino actually has a tiny mass. The experiment, as I have explained, hinges on whether the maximum electron energy is less than E^*. (Recently, a Russian group announced that the neutrino mass is not zero. Other experimental groups, however, have failed thus far to corroborate this finding.)

Chapter 4: Marriage of Time and Space

- Page 52. The poem by Y. Chen is from *An Annotated Anthology of Sung Dynasty Poems* [in Chinese], edited by Zhung-shu Chien (Beijing: People's Literature Publishers, 1979), p. 148. Translation in the text is by this author.
- Page 56. For the flashback on the development of electromagnetic the-

ory, I have consulted the following historical sources: B. Morgan, *Men and Discoveries in Electricity* (London: Wyman and Sons, 1952); G. Holton and D. H. D. Roller, *Foundations of Modern Physical Science* (Reading, Mass.: Addison-Wesley, 1958); J. C. Maxwell, *Physical Thought from the Pre-Socratics to the Quantum Physicists,* an anthology edited by S. Sambursky (New York: Pica Press, 1975).

- Page 56. The word "electric" comes from the Greek word for amber.
- Page 56. Children are unfailingly fascinated by magnets. Here is a force that operates even between two bodies that are not in contact with each other. To account for this phenomenon, the Roman poet Lucretius proposed that "a shoal of seeds" streams forth from a lodestone, "driving away with its blows" the surrounding air. This forms a vacuum into which "the atoms of iron tumble headlong." (While this amusing theory is wrong, Lucretius showed a remarkable understanding of the effects of a vacuum.)
- Page 57. In 1600, Gilbert published his research in one of the most influential books in the history of physics. Indeed, the Inquisition charged Galileo with owning a copy of Gilbert's *De magnete,* among other misdeeds. It is also noteworthy, incidentally, that the English navy, under Charles Howard, defeated the Spanish Armada in 1588. It was neither the first nor the last time that commercial and military interests had stimulated research.
- Page 58. The question as to whether the passage of an electric current through a wire in the vicinity of a magnet would cause the wire to move was asked by William Wollaston (1766–1828), now long forgotten by all but historians of science.
- Page 65. In 1887, when Einstein was eight years old, American physicists Albert Michelson and Edward Morley, stimulated by an inquiry that Maxwell addressed to the U.S. Navy, carried out an ingenious experiment to determine the speed of light, as measured by observers in relative uniform motion. Textbooks often present this experiment as one of the most crucial in the development of physics, and, indeed, the surprising result found by these two caused quite a stir in the physics community. Einstein, however, gave no indication whatsoever in his work that he was aware of the Michelson–Morley experiment. As I have explained, he could well have reached the same conclusion as they by purely theoretical reasoning. The question as to whether Einstein knew of the experiment before 1905 has long intrigued science historians. Abraham Pais, the leading biographer of Einstein, sifted the historical records and concluded that yes, Einstein did know. See A. Pais, *Subtle Is the Lord: The Science and the Life of Albert Einstein* (New York: Oxford University Press, 1982).
- Page 65. Another bit of history involves the talented Frenchman Armand Hippolyte Louis Fizeau. In 1851, he measured the speed of light

in moving streams of water, but the result of his measurements sat unexplained for half a century until Einstein's theory came along.

- Page 68. In 1928, Einstein suggested to Piaget that he should study how children perceive time. See J. Piaget, *The Child's Conception of Time* (New York: Ballantine, 1971), p. vii.
- Page 68. The author of the limerick about Miss Bright is an expert on fungi. It can be found in W. S. Baring-Gould, *The Lure of the Limerick: An Uninhibited History* (New York: C. N. Potter, 1967).
- Page 68. Magellan's voyage is recounted in S. Zweig, *The Story of Magellan* (Philadelphia: Century Book Bindery, 1983). Incidentally, Lewis Carroll was among those who, in 1878, proposed the introduction of international time zones to resolve the paradox of how Magellan lost a day.
- Page 68. In accounts such as this, and, for that matter, in physics texts as well, one naturally tends to attribute any given development to an individual physicist. One does not, by necessity, clutter up the text with the false leads and blind alleys so prevalent in science; for that, look to a specialized historical treatise.

  By giving a long flashback on electromagnetism, I hope to convey to the reader a sense of the Zeitgeist prevalent in the physics community around the turn of the century. The point is, electromagnetic theory had developed to the point that the relativistic-invariance issue came naturally into the consciousness of Einstein's generation. Art historians are known to use the same kind of argument to explain why, for example, the baroque influence surfaced more or less simultaneously in different countries. Similarly, a number of physicists besides Einstein—Henri Poincare in France, Henrich Lorentz in Holland, George Fitzgerald in England, Woldemar Voigt and Herman Minkowski in Germany—all had wrestled with relativity. Einstein, however, went the furthest in extracting the physical consequences of relativistic invariance.
- Page 71. For those of you who want to know more about Einstein's revision of mechanics, here are some salient points. A central notion in mechanics, of course, is that the velocity of a moving object is defined as the distance the object traveled divided by the time elapsed. But which time?

  Should we use the object's proper time, or the time clocked by an observer watching the object go by? Physicists betray their prejudices and call these two possible definitions velocity and improper velocity, respectively. In everyday experiences, the distinction is entirely negligible. For fast moving objects, however, the proper and improper velocities can differ enormously.

  The photon provides the most extreme example. Recall, the clock carried by a photon is stuck perpetually at high noon. The proper time of the photon does not change. Thus, the proper velocity of light is actually infinite. The improper velocity of light, in contrast, is perfectly finite, equal to about 300,000 km per second. Some laymen find one aspect of relativity particularly fascinating: the existence of an ultimate speed limit imposed by the speed of light. Actually, this speed limit refers to the improper velocity of light, not the proper.

In mechanics, the momentum of a moving object is equal to its mass times its velocity, an eminently reasonable formula. The moving Mack® truck has more momentum than the passenger sedan moving alongside it. Einstein, in considering the motion of fast moving objects, had to decide whether the definition of momentum should involve the proper or the improper velocity.

The choice of which definition to use in physics is influenced by the desire that the equations look as "clean" and as symmetric as possible. Physicists, in essence, require physical quantities to *transform* neatly under the relevant transformations, the Lorentz in this case. Proper velocity wins this contest hands down. Under a Lorentz transformation, the denominator in the definition of proper velocity—the proper time of the moving object—does not change at all. In other words, the proper time of a moving object by definition is an intrinsic property of the moving object and does not depend on the observer. The elapsed time clocked by an observer, on the other hand, depends on the observer. Einstein chose to define momentum using the proper velocity, a choice that inexorably drove him to his formula, $E = mc^2$: Once the definition of momentum is fixed, the definition of energy follows, since, under Lorentz transformation energy and momentum are related.

Another important consideration is that momentum is conserved if it is defined in terms of the proper velocity rather than the improper. (See Chapter 8 for a discussion of conservation laws and their close relationship to symmetry considerations.)

It logically follows from the formula for momentum—the momentum of an object is equal to its mass times its proper velocity—that a massive object cannot move at the speed of light. If it did, it would have infinite proper velocity and, hence, infinite momentum. Since momentum, as opposed to proper velocity, is a measurable physical quantity, an object is not allowed to carry infinite momentum. Conversely, a massless particle, such as the photon or neutrino, must travel at the speed of light in order to carry any momentum at all.

Chapter 5: A Happy Thought

- Page 76. Einstein expressed his dislike for the word "relativity" in a letter to E. Zschimmer, September 30, 1921 (as quoted by G. Holton, ibid.).
- Page 76. See Lawrence Durrell, *Balthazar* (New York: Dutton, 1958), pp. 9, 142.
- Page 76. In his article, "Introduction: Einstein and the Shaping of Our Imagination," published in *Albert Einstein, Historical and Cultural Perspectives,* edited by G. Holton and Y. Elkana (Princeton, N.J.: Princeton University Press, 1982), Holton lists and analyzes the enormous misconceptions of Einstein's work that have become part of our culture. In his estimation, one writer who successfully incorporated Einstein's thoughts is William Faulkner (in *The Sound and the Fury*). Holton writes that in Faulkner "it is futile to judge whether the traces of modern physics are good physics or bad, for these trace elements have been used in the making of a new alloy."
- Page 83. S. Weinberg, *Gravitation and Cosmology* (New York: Wiley, 1972).
- Page 84. Peters map, by A. Peters, is published in the United States by Friendship Press; a description can be found in the April 1984 issue of *Harper's*. Thanks to Peggy Gallagher for calling my attention to this reference.
- Page 86. Incidentally, the notion that gravity could bend light occurred to Newton, who thought that light consisted of tiny, dense balls.
- Page 89. In 1884, Harvard University erected the first building in the United States devoted exclusively to physics and lured Edwin P. Hall, an early American experimentalist of some note, away from Johns Hopkins. He had taken some important electromagnetic measurements, but now he became enthusiastic over the notion of detecting possible deviations from Newton's law of free fall. To accommodate him, Harvard incorporated a tower some twenty meters high in the new building. Hall, naturally, did not discover any interesting effects, and three quarters of a century passed before the tower was put to sensible use.
- Page 90. To truly understand black holes, we must grasp an essential difference between Newton and Einstein.

    In Newton's theory, a massive object generates a gravitational field around it, and that is the end of the story. The situation with Einstein is considerably more complicated. Physicists have understood that a given field contains energy since Maxwell's time; consider the gravitational field around a star, for instance. Since, according to Einstein's earlier work, energy is equivalent to mass, that gravitational field contains mass and, thus, generates an additional gravitational field. The star, in other words, generates a gravitational field, which, in turn, generates an additional field, which generates yet another field, and so on, ad infinitum. It is the gravitational analogy of money begetting more money through compound interest. In this way, Einstein's theory exemplifies a so-called nonlinear theory; Newton's theory is linear. Normally, the additional gravitational fields generated are small, and Einstein's theory differs little from Newton's. Around a black hole, however, the additional gravitational fields pile up, causing an extreme warping of spacetime.

- Page 92. Hubble built on the earlier work of Vesto Slipher, as well as that of his colleague Milton Humason. See H. Pagels, *Perfect Symmetry* (New York: Simon & Schuster, 1985).
- Page 93. The comparison between Einstein's work and Beethoven's is taken from A. Pais, *Subtle Is the Lord: The Science and the Life of Albert Einstein* (New York: Oxford University Press, 1982). Pais remarks on the aptness of the motto carried by Beethoven's Opus 135.

Chapter 6: Symmetry Dictates Design

- Page 96–97. The figures on these pages are adapted from C. N. Yang, "Einstein and His Impact on the Physics of the Second Half of the Twentieth Century," CERN report, 1979.
- Page 98. Einstein's theory looks quite simple. So does Newton's when expressed in terms of what Newton called the gravitational field. But Einstein's theory expressed in Newton's field becomes a ghastly mess, consisting of an infinite number of terms. It is a safe bet that no one could have guessed at this infinite series without the help of a symmetry principle.

Chapter 7: Where the Action Is Not

- Page 107. The mathematics developed in connection with the action principle has spread from physics to a variety of other fields in which strategic planning plays a role. The runner wants to determine the history that would minimize the time it takes to finish a race. Indeed, just such an analysis of racing has appeared in a physics journal. The relevance of economics has already been suggested in our discussion. Eminent economist Paul Samuelson began his Nobel Prize lecture, "Maximum Principles in Analytic Mechanics," delivered December 11, 1970, with a discussion of the action principle in physics. He went on to speak about a profit-maximizing firm whose output is controlled by ninety-nine different inputs; he wrote:

> [an] economist could in principle record . . . ninety-nine demand functions relating the quantity of each input bought by the firm to the

ninety-nine variables depicting the input prices. . . . What a colossal task it would be to store bits of information defining ninety-nine distinct surfaces in a one hundred dimensional space! [Modern economics is made manageable by the recognition that] the observed demand curves . . . are actually themselves solutions to a maximum-profit problem.

- Page 109. Some people have fantasized that the least-action principle can apply to the nonphysical world as well as the physical, imagining that somewhere there is a film library containing all possible histories. In one history, Romulus and Remus are eaten, rather than nourished, by the she-wolf. In another, Napoleon defeats Wellington. Each history is assigned a number. Out of this infinity of histories, the one with the smallest number is chosen!

Chapter 8: The Lady and the Tyger

- Page 114. I have taken the facts of Noether's life from her primary biography by A. Dick, *Emmy Noether 1882–1935,* English translation by H. I. Blocher (Boston: Birkhauser, 1981), and from *Emmy Noether: A Tribute to Her Life and Work,* edited by J. W. Brewer and M. K. Smith (New York: Marcel Dekker, 1981), as well as from the memorial addresses given by her contemporaries, many reprinted in the two works just cited. Brewer and Smith's book appeared in the series *Monographs and Textbooks in Pure and Applied Mathematics,* and while much of it is of a purely mathematical nature, it does contain biographical material.
- Page 118. According to E. Wigner, in an article published in *Understanding the Fundamental Constituents of Matter,* edited by A. Zichichi (New York: Plenum, 1978), J. F. C. Hessel, in studying the symmetries of crystals, was the first physicist to discuss symmetries explicitly. One of the first systematic discussions was given by A. Kretschman, *Ann. der Physik* 53 (1917), p. 575. The connection between invariance and conservation was studied with varying degrees of generality by G. Hamel, *Z. Math. Phys.* 50 (1904), p. 1; E. Noether, *Nach. Ges. Wiss. Göttingen* (1918), p. 235; and F. Engel, *ibid.,* p. 375.
- Page 121. In April 1933, one month after the inception of the Third Reich, Noether, who was of Jewish descent, was expelled from the university. She died two years later at Bryn Mawr College.

Chapter 9: Learning to Read the Great Book

- Page 125. Of all the math courses I ever took, group theory was the most fun.

- Page 127. A note to the mathematicians who may happen to read *Fearful Symmetry:* I discuss the rotation group, and not its covering group.

Chapter 10: Symmetry Triumphs

- Page 140. Heisenberg's uncertainty principle has suffered the same fate as Einstein's principle of relativity at the hands of writers to whom the word "uncertainty" conjures up images never envisioned by Heisenberg. I learned, to my astonishment, that Heisenberg's uncertainty principle is bandied about in some architectural circles. Others have stretched it to mean whatever they want it to mean. I will refrain from denouncing this phenomenon further; instead, I once again quote economist Paul Samuelson, who, in his Nobel lecture, stated:

  > There is really nothing more pathetic than to have an economist or a retired engineer try to force analogies between the concepts of physics and the concepts of economics. . . . when an economist makes reference to a Heisenberg Principle of Indeterminacy in the social world, at best this must be regarded as a figure of speech or a play on words, rather than a valid application of the relations of quantum mechanics.

- Page 142. In quantum physics, probability amplitudes are not given by ordinary numbers, but by what mathematicians call complex numbers.
- Page 143. In the 1960s, Steve Adler, and, independently, John Bell and Roman Jackiw, discovered that, under certain circumstances, a quantum theory may not have all the symmetries possessed by the corresponding classical theory. At that time, this theoretical possibility was so unexpected that it became known as the anomaly. The study of anomalies now plays an important role in our search for the symmetries of nature.
- Page 146. I have simplified the discussion of the role of rotational symmetry in atoms by neglecting the electron's spin.

Chapter 11: The Eightfold Path in the Forest of the Night

- Page 159. Here is a brief outline of the history of isospin symmetry. Since the neutron is so close in mass to the proton, Chadwick naturally assumed that the neutron consisted of a proton with an electron stuck on it. Thus, atomic nuclei were thought to consist of protons and electrons. We now know that this picture is incorrect and contradicts a number of experimental observations. It was difficult to understand how some of the electrons in the atom managed to orbit around the nucleus while others were sucked into the nucleus.

Heisenberg proposed that the neutron is a particle in its own right and that the nucleus consists of protons and neutrons. He then supposed that strong interaction physics remains invariant if one *exchanges* the proton and the neutron. Note that this symmetry is considerably weaker than isospin symmetry, in which one transforms the proton and the neutron into linear combinations of each other. Heisenberg, however, continued to think of the neutron as a proton with an electron attached. He explained the origin of the interaction between the proton and the neutron as follows: When a neutron gets close to a proton, the electron inside the neutron may hop over to the proton. Heisenberg reasoned that the electron, by hopping back and forth between the proton and neutron, could produce an interaction between the two. In Heisenberg's picture, there is no strong interaction between two protons, since there is no electron around to hop back and forth. The atomic nucleus was erroneously supposed to be held together by the attraction between protons and neutrons.

Heisenberg's theory was proven wrong by the experimentalists N. P. Hydenburg, L. R. Hafstad, and M. Tuve, who measured the strong interaction between two protons (following earlier work of M. White) and discovered it to be comparable in strength to the interaction between a proton and a neutron. In 1936, B. Cassen and E. U. Condon, and, independently, G. Breit and E. Feenberg, proposed that Heisenberg's exchange symmetry be generalized to isospin symmetry. (I thank S. Weinberg for a helpful discussion on this point.)

Incidentally, only the electrically charged pions appeared in Yukawa's 1934 paper. The fact that isospin requires the electrically neutral pion was not recognized until considerably later (around 1938) by N. Kemmer, and, independently, by S. Sakata, M. Taketani, and H. Yukawa.

- Page 166. I have quoted from Hideki Yukawa's autobiography, *Tabibito* (*The Traveler*), English translation by L. Brown and R. Yoshida (Singapore: World Scientific Publishing, 1982). In it, Yukawa describes the "long days of suffering" he endured from 1932 to 1934 searching for a theory of the nuclear force. To calm himself, he tried sleeping in a different room every night. The crucial point came to him in a flash one night in October 1934.

- Page 167. I simplified the discussion of strangeness conservation by saying that the K^0 is always produced with a Σ^+. Also, the route by which Murray Gell-Mann, Kazuo Nishijima, Abraham Pais, and others arrived at strangeness conservation was considerably more arduous than what the text might suggest.

- Page 168. The quotation attributed to E. Fermi appears in *More Random Walks in Science,* compiled by R. L. Weber (Bristol, England: The Institute of Physics, 1982).

- Page 168. Incidentally, there's a limit to linguistic purity among physicists. The plural "leptons" is used instead of the Greek *lepta*.
- Page 168. The term "mesoton" was assigned by experimenters Carl Anderson and Seth Neddermeyer to a particle that they had discovered. Physicist Robert Millikan suggested changing it to "mesotron" in order to be consistent with the terms electron and neutron, but, as Anderson remarked, not proton. The awful mesotron was used and later shortened to "meson," at the suggestion of the Indian physicist Homi Bhabha. According to George Gamow, some French physicists protested, fearing confusion with their word for house. Meson has the same sound as the word 迷想, Chinese and Japanese for hallucination or illusion. In the 1930s, Japanese physicists met regularly to discuss meson physics in what were known as illusion meetings.

  It turned out that the particle discovered by Anderson and Neddermeyer is not Yukawa's. To distinguish between them, Yukawa's particle was called the pi meson, and the impostor, the mu meson. Later, it was realized that the mu meson is not a meson at all; it is a lepton, like the electron, and its name was shortened to muon (see Chapter 15). Incidentally, Yukawa in his paper referred to the pion as the U-particle. See the article by Carl D. Anderson in *The Birth of Particle Physics,* edited by Laurie Brown and Lillian Hoddeson (New York: Cambridge University Press, 1983), p. 148. I thank S. Hayakawa and L. Brown for a helpful discussion of the history associated with the name meson.
- Page 172. As an indication of the confused experimental situation in the early 1960s, consider the fact that of twenty-six hadrons (with certain properties) listed in a survey published in 1963, nineteen are now known not to exist.
- Page 173. The reader may note that the group $SU(3)$ and the rotation group $SO(3)$ both involve transformations of three objects in their defining representations. The transformations involved are different for the two groups, however.
- Page 174. I have consulted Yuval Neéman's autobiographical article, "Hadron Symmetry, Classification, and Compositeness" (to be published), for material on his life. See also R. Deacon, *The Israeli Secret Service* (New York: Taplinger, 1977), p. 318.

  One learns from this quasi-autobiographical account that Neéman's ancestors were among the disciples of Rabbi Eliyahu the Goan (1720–1797), a group which represented the cutting edge of rationalistic and scholarly Judaism, the "resisters" against the "sentimentalist" Hassidic view.

  According to Neéman, in an article published in *The Interaction Between Science and Philosophy,* edited by Y. Elkana (Atlantic Heights, N.J.: Humanities Press, 1974), pp. 1–26, Sakata, who was a

.confirmed Marxist, was led astray by his insistence on a philosophy of Nature based on dialectical materialism. I thank Yuval Neéman for sending me copies of his writings.

- Page 177. Gell-Mann referred to haute cuisine in his paper Phys. *l* (1964), p. 63.
- Page 179. A personal account of the invention of quarks may be found in M. Gell-Mann, "Particle Theory from S-Matrix to Quarks," a talk presented at the First International Congress on the History of Scientific Ideas at Sant Feliu de Guixols, Catalunya, Spain, in September 1983.
- Page 179. Given that the proton is made of two up quarks and a down quark, and that the neutron is made of two down quarks and an up, we can easily work out the electric charges of the quarks. The electric charges of the proton and of the neutron are given simply by the sum of the electric charges of the quarks they contain. We can change a neutron into a proton by changing one of its down quarks to an up. Recall, the proton carries one unit of electric charge while the neutron carries none. The up quark, therefore, must carry one more unit of charge than the down. Denote the up quark's charge by Q. Then the down carries the charge $Q - 1$. The proton, containing two up quarks and a down, would then have charge $Q + Q + Q - 1 = 3Q - 1$. Since the proton has charge 1, we obtain the equation $3Q - 1 = 1$, which we can see is solved by $Q = \frac{2}{3}$. Some of the most important calculations in fundamental physics are not that difficult.

   That quarks have fractional charges disturbed many physicists and accounted for Gell-Mann's initial reluctance to consider the defining triplet representation of $SU(3)$. Until then, all known particles had charges equal to multiples of the proton charge.

Chapter 12. The Revenge of Art

- Page 185. The remarkable discovery that James Joyce had known about gauge symmetry back in 1914 was made by Predrag Cvitanovic, *Field Theory*, Nordita lecture notes (Copenhagen: Nordita, Blegdamsvej, 1983), p. 72. I thank William Bialek for showing me this book.
- Page 185. I cannot resist reproducing the passage in which Joyce spoke of gauge symmetry. I am sure that many readers would be curious about the context. Here it is.

   Zoe: (Stiffly, her finger in her neckfillet.) Honest? Till the next time. (She sneers.) Suppose you got up the wrong side of the bed or came too quick with your best girl. O, I can read your thoughts.
   Bloom: (Bitterly.) Man and woman, love, what is it? A cork and bottle.

Zoe: (In sudden sulks.) I hate a rotter that's insincere. Give a bleeding whore a chance.

Bloom: (Repentantly.) I am very disagreeable. You are a necessary evil. Where are you from? London?

Zoe: (Glibly.) Hog's Norton where the pigs play the organs. I'm Yorkshire born. (She holds his hand which is feeling for her nipple.) I say, Tommy Tittlemouse. Stop that and begin worse. Have you cash for a short time? Ten shillings?

Bloom: (Smiles, nods slowly.) More, houri, more.

Zoe: And more's mother? (She pats him offhandedly with velvet paws.) Are you coming into the musicroom to see our new pianola? Come and I'll peel off.

Bloom: (Feeling his occiput dubiously with the unparalleled embarrassment of a harassed pedlar gauging the symmetry of her peeled pears.) Somebody would be dreadfully jealous if she knew. The greeneyed monster. (Earnestly.) You know how difficult it is. I needn't tell you.

Zoe: (Flattered.) What the eye can't see the heart can't grieve for. (She pats him.) Come.

Bloom: Laughing witch? The hand that rocks the cradle.

Zoe: Babby!

Bloom: (In babylinen and pelisse, bigheaded, with a caul of dark hair, fixes big eyes on her fluid slip and counts its bronze buckles with a chubby finger, his moist tongue lolling and lisping.) One two tlee: tlee tlwo tlone.

- Page 187. Ideas similar to that of Yang and Mills had also been discussed by O. Klein and R. Shaw.
- Page 188. Niels Hendrik Abel was born the son of a poor pastor in rural Norway. A brilliant mathematician, he died in abject poverty at the age of twenty-six. Remember that the order in which one multiplies together two transformations in a group makes a difference. Abel's name is associated with groups in which the order of multiplying transformations makes no difference; they are known as abelian. Physicists generally are interested in non-abelian groups, those in which the order of multiplying transformations does make a difference; hence, the term non–abelian gauge theories. Our theory of the strong interaction, for example, is a non–abelian gauge theory. It is ironical that this brilliant mathematician's name is now routinely invoked by physicists negatively. Incidentally, electromagnetism is an example of an abelian gauge theory.
- Page 191. The information I gave about the word "gauge" was gleaned from several dictionaries. *The Oxford English Dictionary* states, however, that the word is of unknown origin. It appeared first in Old Northern French in the thirteenth century, but it is left wanting in other romance languages. (In modern French, gauge is spelled jauge.) Weekley concurs with the *OED* but includes a questionable etymological attribution to Middle High German; see E. Weekley, *An Etymological*

Dictionary of Modern English (New York: Dover, 1967). In any case, gauge theories in physics have nothing to do with distance or size.

- Page 197. The notion that coupling strengths may vary with the energy scale at which the physical world is examined was pointed out in the early 1950s by E. C. G. Stueckelberg and A. Peterman, by M. Gell-Mann and F. Low, and by N. Bogolyubov and D. V. Shirkov. The physics community on the whole did not pursue this idea partly because the relevant papers were very difficult to read.
- Page 199. I called asymptotically free theories "stagnant theories," since once the coupling strength has moved to zero, it stays there.
- Page 201. G. 'tHooft also had realized independently that Yang–Mills theory is asymptotically free. His findings, however, were not published in a journal and not widely known. Soviet physicist I. B. Khriplovich had also been studying the behavior of the coupling strength in Yang–Mills theory.
- Page 201. Incidentally, my work with Wilczek on the neutrino scattering off protons marked the beginning of a long collaboration. We spent a number of years together at Princeton, where our growing families got to know each other very well. Wilczek and I both are now at the Institute for Theoretical Physics in Santa Barbara.
- Page 208. That Yang–Mills theory is renormalizable was also discussed by B. W. Lee and J. Zinn-Justin.

Chapter 13: The Ultimate Design Problem

- Page 212. Thomas Mann, *The Magic Mountain*, translated by Helen Lowe-Porter (New York: Knopf, 1927), p. 480.
- Page 214. The notion of spontaneous symmetry breaking came to fundamental physics by a rather circuitous route. The examples we gave, involving the wine bottle and the magnet, indicate that many common physical phenomena exhibit spontaneous symmetry breaking, a fact that was often not specifically recognized. In 1957, three American physicists, John Bardeen, Leon Cooper, and Robert Schrieffer, managed to explain a peculiar phenomenon known since the early 1900s, the phenomenon of superconductivity. As is well known, a metal wire displays a certain resistance to the passage of electric current. But when various metals are cooled to very low temperatures, this resistance dramatically disappears. Bardeen, Cooper, and Schrieffer were awarded the Nobel prize for their work. (Incidentally, John Bardeen thereby became the only person in history to have won two Nobels in the same field, having won it earlier with the discovery of the transistor.) I will refrain from giving a detailed account of how superconductivity is explained, but suffice to say that it involves spontaneous symmetry breaking in an essential way. This is particularly evident in a

version of the theory by Russian physicists V. Ginsburg and Lev Landau. Many physicists, notably Stephen Adler, Curtis Callan, G. Jona-Lasinio, Maurice Lévy, Murray Gell-Mann, Murph Goldberger, Sam Treiman, and William Weisberger, besides the ones mentioned in the text, contributed to our understanding of spontaneous symmetry breaking.

- Page 218. Gell-Mann's "veal-flavored pheasant" symmetries turned out to be both explicitly and spontaneously broken. Thus, the task of bringing these symmetries to light was extremely arduous.

Chapter 14: Unity of Forces

- Page 234. Some reader may wonder why the fifteen quark and lepton fields cannot be assigned to three copies of the five-dimensional representation. They can, but then the quarks and leptons would not have their observed properties. That is what I meant when I said in the text that the fit is more seamless than our simple counting would suggest. The point is that, a priori, there is no guarantee that the quarks and leptons would come out with their observed properties. (Also, were one to use this assignment, the theory would suffer from the anomaly. See note to Chapter 10.)
- Page 234. Naturally, some wag has referred to the assignment of quarks and leptons in the $SU(5)$ theory as the "Woolworth assignment."
- Page 236. To streamline the presentation, I have taken some liberties and explained grand unification in terms of an "electromagnetic hiker" and a "weak hiker." Strictly speaking, I should say a "$SU(2)$ hiker" (who comes down) and a "$U(1)$ hiker" (who climbs up).
- Page 236. The presence of three hikers rather than two may puzzle some readers. Even though the electroweak interaction unifies the electromagnetic and the weak interactions, it still contains two different coupling strengths. See also the preceding note.
- Page 238. When I mentioned neutron decay, I once again sacrificed strict scientific accuracy. The neutron actually decays into a proton, an electron, and an antineutrino.
- Page 246. In an influential article, Gary Steigman examined the available evidence and concluded that the matter-antimatter universe is not tenable.
- Page 252. Previously, I authored a book on grand unification, *Unity of Forces in the Universe*, 2 vol. (Singapore: World Scientific Publishing, 1982). While it is addressed to graduate students and researchers in this field, the lay reader who wants to know more about grand unification can get a flavor of what some of the ongoing research is like from glancing through it.

- Page 252. Long before grand unification, the great Soviet physicist and humanitarian Andrei Sakharov had speculated that the origin of matter in the universe may be understood if the number of baryons is not strictly conserved. Few in the West were aware of Sakharov's work. After grand unification was proposed, a number of physicists, including M. Yoshimura, S. Dimopoulos, L. Susskind, D. Toussaint, S. Treiman, F. Wilczek, S. Weinberg, and myself, reinvented the scenario for the genesis of matter and placed it in the framework of grand unification. My own involvement with this problem dates back to the early 1970s, when I went for an extended visit to Paris. I rented an apartment belonging to a collaborator of Roland Omnès, a French physicist who had advocated a universe filled equally with matter and antimatter. There, I came across various papers dealing with this scenario. Later, together with Frank Wilczek, I tried on and off for a couple of years to see if spontaneous symmetry breaking might provide a segregation mechanism in the matter–antimatter universe, but to no avail. Finally, working with Toussaint and Treiman, we realized that baryon nonconservation provided the key.
- Page 252. Of the many difficulties facing the inflationary universe idea, economists will appreciate the "graceful exit problem": getting the universe out of the inflationary epoch. To solve it, A. Linde, A. Albrecht, and P. Steinhardt invented variants of the original model.

Chapter 15: The Rise of Hubris

- Page 255. The fact that fundamental physicists have changed the questions reminds me of an old academic joke about a graduate student in economics getting ready for his doctoral exam. The student decided to prepare by looking up the exams from the preceding years. To his surprise, he found that the same questions were asked year after year. Confronting his professor about this fact, the graduate student was told by the distinguished academic, "Oh yes, the questions are the same every year, but the correct answers change from year to year."
- Page 256. The muon was also discovered independently by Jabez Street and E. C. Stevenson.
- Page 256. The confusion over the muon was cleared up in the 1940s by S. Sakata and T. Inoue in Japan, and by R. Marshak and H. Bethe in the United States.
- Page 261. The apartness of Einstein was described by Pais in his biographical treatise (particularly page 39).
- Page 264. Higher dimensional theories of gravity were also proposed independently by Gunnar Nordstrom and by Heinrich Mandel (see Pais).

- Page 269. The notion that the grand unified gauge interaction may beget gravity originated in the work by Soviet physicist and human rights activist Andrei Sakharov. A number of other physicists, including P. Minkowski, Y. Fujii, H. Terazawa, S. Adler, and myself, rediscovered this idea later and developed it.
- Page 270. Fermions are named in honor of the Italian-American physicist Enrico Fermi, while bosons are named for Indian physicist Satyendra Bose.
- Page 271. Julius Wess and Bruno Zumino, working in Europe, were the first to investigate supersymmetry systematically. Earlier discussions of supersymmetry appeared in the works of Soviet physicists Y. Golfand, E. Likhtman, D. Volkov, and V. Akulov. Indications of supersymmetry also appeared in the work of Andrei Neveu, John Schwarz, and Pierre Ramond.
- Page 274. Einstein, "On the Method of Theoretical Physics," the Herbert Spencer lecture delivered at Oxford, June 10, 1933, published in *Mein Weltbild* (Amsterdam: Querido Verlag, 1934).

Chapter 16: The Mind of the Creator

- Page 277. The *CPT* theorem was discovered by G. Lüders, B. Zumino, W. Pauli, J. Schwinger, and others.
- Page 277. A priori, it is logically possible that *CPT* invariance, rather than *T* invariance, is violated in *K* meson decay, but a careful analysis of the experimental evidence indicates that *CPT* invariance is respected.
- Page 277. We are all familiar with how spinning tops precess in gravitational fields. Searching for direct evidence of violation of time reversal invariance, experimenters are studying the precession of various particles, such as the electron and the neutron, in electric fields. The precession picks out a direction.
- Page 278. It has also been suggested that the arrow of time in our consciousness is linked to the expansion of the universe. But, it is difficult to see how the movement of faraway galaxies could possibly affect the working of our consciousness.
- Page 278. For a sampling of writings on the nature of time, see *The Enigma of Time*, compiled and introduced by P. T. Landsberg (Bristol, England: Adam Hilger, 1982).
- Page 280. I have barely touched, of course, on the mystery of the human consciousness. For an introduction to the literature, see C. Hampden-Turner, *Maps of the Mind* (New York: Macmillan, 1981). A clear and popular introduction to quantum measurement theory has been given by H. Pagels in *The Cosmic Code* (New York: Simon & Schuster, 1982). Writing in the professional physics literature, H. D.

Zeh of the University of Heidelberg, and A. Leggett of the University of Illinois, are among those who in recent years have produced insightful analysis on the relation between the observer and the quantum.

- Page 280. For a critique of Blake's poem, see Lionel Trilling, *The Experience of Literature* (Garden City, N.Y.: Doubleday, 1967), p. 857. The reader may recall that the last stanza of Blake's poem on the Burning Tyger is the same as the first, except that the word "could" in the line "Could frame thy fearful symmetry?" is changed to "dare" in the last stanza. Interestingly enough, in earlier drafts, Blake had the same first and last stanzas, using "dare" in both. We should of course refrain from reading too much of the poem into the search for symmetries in contemporary physics.
- Page 281. The quotation on whether or not God had any choice was a remark made by Albert Einstein to Ernst Straus, and it can be found in *Einstein: A Centenary Volume,* edited by A. P. French (Cambridge, Mass.: Harvard University Press, 1979). The quotation by B. Hoffmann may be found in his *Albert Einstein: Creator and Rebel* (New York: Viking, 1972).
- Page 283. The anthropic argument has been severely criticized over the years. For a reasoned critique of the anthropic arguments, see the article by H. Pagels in *The Sciences,* vol. 25, no. 2, 1985.

Afterword

- Page 285. For a brief yet substantive introduction to string theory by the acknowledged master of the subject, see Ed Witten, "Reflections on the Fate of Spacetime," *Physics Today* (April 1996), p. 24. For a full treatment, the best book available is J. Polchinski, *Superstrings* (Cambridge: Cambridge University Press 1998).
- Page 287. The recent development showing that the consistency of superstring theory requires the presence of the p-branes was spearheaded by my colleague at the Institute for Theoretical Physics, Joseph Polchinski, and his junior collaborators, Jin Dai and Rob Leigh, building upon earlier work by a number of physicists. For references to earlier work, see J. Polchinski, TASI lectures on D-branes (available on the World Wide Web).
- Page 292. A. Zee, "The Unreasonable Effectiveness of Symmetry in Fundamental Physics," in *Mathematics and Science*, edited by Ronald E. Mickens (Teaneck, N.J.: World Scientific, 1990).
- Page 294. A. Zee, "On Fat Deposits around the Mammary Glands on

the Females of Homosapiens," *New Literary History*, ed. Ralph Cohen, forthcoming 2000.

- Page 294. Richard Dawkins, *The Blind Watchmaker* (New York: W.W. Norton & Co., 1986).
- Page 294. George C. Williams, *The Pony Fish's Glow* (New York: Basic Books, 1997), p. 140.
- Page 294. Lee Smolin, *The Life of the Cosmos* (Oxford: Oxford University Press, 1998).

Index